THE
METRIC SYSTEM
MADE SIMPLE

THE METRIC SYSTEM MADE SIMPLE

By ALBERT F. KEMPF
and THOMAS J. RICHARDS

MADE SIMPLE BOOKS
DOUBLEDAY & COMPANY, INC.
GARDEN CITY, NEW YORK

Material in this book has been taken from
Using the Metric System and *Exploring the Metric System*
By Albert F. Kempf and Thomas J. Richards
Published by Laidlaw Brothers

Copyright © 1973
LAIDLAW BROTHERS · PUBLISHERS
A Division of Doubleday & Company, Inc.
River Forest, Illinois
Palo Alto, California Dallas, Texas Atlanta, Georgia Toronto, Canada

ISBN: 0-385-11032-4
Library of Congress Catalog Card Number 75–36631
Printed in the United States of America

Library of Congress Cataloging in Publication Data

Kempf, Albert F.
The metric system made simple.

(Made simple books)
"Material . . . taken from [the authors'] Using the
metric system and Exploring the metric system."
1. Metric system.   2. Weights and measures—United
States.   I. Richards, Thomas J., joint author.   II. Title.
QC92.U54K45   530′.8

Contents

Introduction

Inches, yards, miles, pounds, tons, pints, gallons. All these terms are familiar American units of measurement whose meanings and equivalencies were drummed into most of us during our elementary school years. In order to solve problems involving computations of length, area, volume, capacity, mass, and weight, we memorized that there are 12 inches in a foot, 3 feet to a yard, 16 ounces to a pound, 4 quarts to a gallon, and so forth.

These varied conversion rates make it time-consuming, if not difficult, to master such a system. If a rod equals 5½ yards and if 40 rods equal a furlong, how many yards are there in 3 furlongs, 2 rods? Obviously, the solution requires several conversions and could hardly be called "simple."

Most American measurements have been based on the English customary unit system since the late 1700s, when congressional action on standardization was taken, although farsighted leaders like Thomas Jefferson opposed the decision. Our yard measurement, for example, was originally the distance from the end of Henry II's nose to the tip of his outstretched hand—not exactly a universal measurement.

Today, even the National Council of Teachers of Mathematics has declared our traditional measurement system unwieldly. Found in its *20th Yearbook* was this observation: "From the point of view of teaching and learning, it would not be easy to design a more difficult system than the present English system of measurement."

Fortunately, in our monetary system, all units of measurement are multiples of 100 or fractions thereof. Calculations require only the addition of zeros or moving the decimal point to the left or right, eliminating complicated fractions and making financial problem-solving easier to teach and learn.

Why aren't all our length, capacity, and weight measurements based on a decimal system? Actually, over 90 per cent of the world's people already use such a system, called the metric system. Based on the *meter,* a unit of measurement universally defined as "the wavelength of orange-red light emitted by a krypton-86 atom," the metric system will soon become the official system of measurement in the United States, the last industrialized nation to make the changeover.

And it's about time we learned to use this method. Yes, it will take time to "unlearn" our present organization of weights and measures, but getting the hang of the metric method will be simple enough. For example: If 1,000 millimeters equal 100 centimeters, and 100 centimeters equal one meter, we see that the ease of unit conversion made possible by decimalization eliminates time-consuming searches for common denominators. In today's supersonic world, "It's always been that way" is no longer an acceptable excuse for our failure to master new and more efficient methods of calculation.

This book is designed to help Americans make the measurement transition effectively through the learn-by-doing method—a time-honored teaching device. Definitions along with exercises involving length, area, volume, and mass calculations are presented, complete with answers for self-checking. The metric system *is* made simple!

Metric standardization will not affect all our measurements. There will still be 60 minutes to an hour and 24 hours to a day. But all present applications of the metric system will be retained, and the jargon once used here exclusively by scientists, pharmacists, medical and technical workers, and the photographic industry will become a basic tool for everyone.

THE
METRIC SYSTEM
MADE SIMPLE

Length

How high?

How long?

Meter

If you live in the United States, you probably use a yardstick to measure your height. But people in most other countries would not use a yardstick. Instead, they would use a meter stick to measure their height. A **meter** is the basic unit of length in all countries that use the metric system.

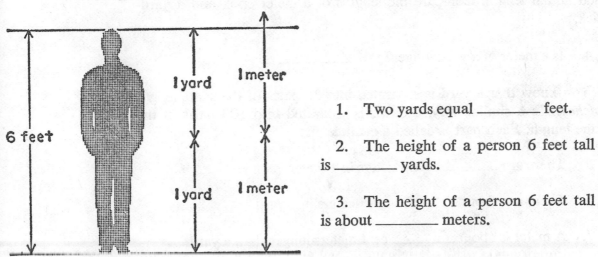

6 feet

1 yard

1 yard

1 meter

1 meter

1. Two yards equal _____ feet.

2. The height of a person 6 feet tall is _____ yards.

3. The height of a person 6 feet tall is about _____ meters.

Note how a meter is defined below. With the proper laboratory equipment, the length of a meter can be determined anywhere in the world, or, for that matter, anywhere in the universe. Consequently, everyone in the world has access to the same standard unit of length.

Meter

A meter is defined in terms of the wavelength of orange-red light emitted by a krypton-86 atom.

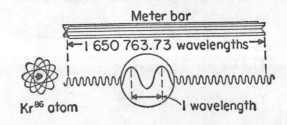

Meter bar

1 650 763.73 wavelengths

Kr⁸⁶ atom

1 wavelength

2   *The Metric System Made Simple*

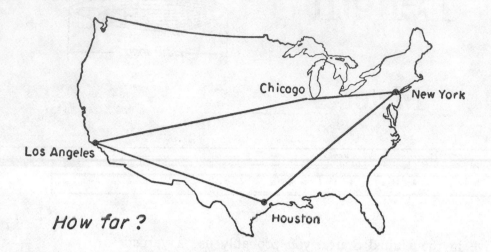

How far ?

The meter and the yard shown on the previous page are one sixth their actual length. Compare the lengths of a meter stick and a yard-stick.

4.   Is a meter or a yard longer? _____

You know that a yard is separated into 36 parts of the same length (*inches*). In a similar way, a meter is separated into 100 parts of the same length. Each part is called a **centimeter.**

5.   There are _____ centimeters in 1 meter.

6.   There are about _____ centimeters in 1 yard.

7.   A meter is about _____ centimeters longer than a yard.

1 meter (m) = 100 centimeters (cm)

Use a meter stick to find the following (to the nearest unit indicated).

8.   The length of your room _____ m

9.   The width of your room _____ m

10.   The length of your desk _____ cm

11.   The width of your desk _____ cm

Decimeter

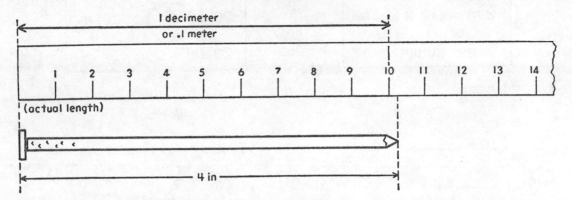

A meter is separated into 10 parts of the same length. Each part is called a **decimeter**. The 4-inch nail is about 1 decimeter long.

1. There are _____ decimeters in 1 meter.

2. There are about _____ inches in 1 decimeter.

1 m = 10 decimeters (dm)   1 dm = .1 m

Cut a strip of paper 1 decimeter long. Use it to find the following.

3. The length of a yardstick   _____ dm

4. The length of a meter stick   _____ dm

5. The length of this page   _____ dm

6. The width of this page   _____ dm

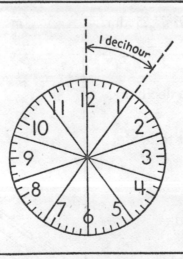

A Decihour

The prefix *deci* means *one tenth of*. Therefore, a deci-hour is one tenth of an hour (.1 × 60) or 6 minutes.

Answer each question.

7. How many seconds are in a deciminute? _____

8. How many hours are in a deciday? _____

9. How many meters are in a decimeter? _____

$$1\ m = 10\ dm$$
$$so$$
$$2\ m = (10 \times 2)\ dm$$
$$or$$
$$2\ m = 20\ dm$$

$$1\ dm = .1\ m$$
$$so$$
$$20\ dm = (.1 \times 20)\ m$$
$$or$$
$$20\ dm = 2\ m$$

10. 1 m = 10 dm
so

1.5 m = (10 × _____) dm

1.5 m = _____ dm

11. 1 m = 10 dm
so

2.63 m = (_____ × _____) dm
or

2.63 m = _____ dm

12. 1 dm = .1 m
so

15 dm = (.1 × _____) m
or

15 dm = _____ m

13. 1 dm = .1 m
so

26.3 dm = (_____ × _____) m
or

26.3 dm = _____ m

Multiplying by 10 moves the decimal point one place to the right
$$10 \times 3.54 = 35.4$$

Multiplying by .1 moves the decimal point one place to the left.
$$.1 \times 3.54 = .354$$

Complete.

14. 6 m = _____ dm   17. 80 dm = _____ m   20. 4.1 m = _____ dm

15. .425 m = _____ dm   18. 6.2 dm = _____ m   21. 29 dm = _____ m

16. 1.2 m = _____ dm   19. .5 dm = _____ m   22. 50 m = _____ dm

A man walked a decimile in a decihour.

23. How many minutes did it take him?

_____

24. How many feet did he walk?

_____

1 mile or 5,280 feet

Centimeter and Millimeter

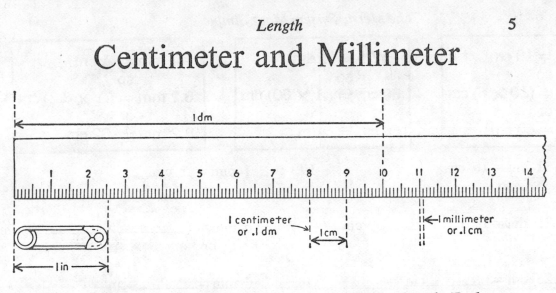

A decimeter is separated into 10 parts of the same length. Each part is called a *centimeter*. Each centimeter is also separated into 10 parts of the same length. Each of these parts is called a **millimeter.**

1.   The 1-inch safety pin is about _____ centimeters long.

2.   The 1-inch safety pin is about _____ millimeters long.

3.   There are _____ centimeters in 1 decimeter.

4.   There are _____ millimeters in 1 centimeter.

$$1 \text{ dm} = 10 \text{ cm} \qquad 1 \text{ cm} = 10 \text{ millimeters (mm)}$$
$$1 \text{ cm} = .1 \text{ dm} \qquad 1 \text{ mm} = .1 \text{ cm}$$

Measure each line segment to the nearest cm, then to the nearest mm.

5.   _____ cm   6.   _____ mm

7.   _____ cm   8.   _____ mm

9.   _____ cm   10.   _____ mm

Measure the following to the nearest cm, then to the nearest mm.

The length of a yardstick   11.   _____ cm   12.   _____ mm

The length of this page   13.   _____ cm   14.   _____ mm

The width of this page   15.   _____ cm   16.   _____ mm

1 dm = 10 cm so 6 dm = (10 × 6) cm or 6 dm = 60 cm	1 cm = .1 dm so 60 cm = (.1 × 60) dm or 60 cm = 6 dm	1 mm = .1 cm so 8.2 mm = (.1 × 8.2) cm or 8.2 mm = .82 cm

17. 1 dm = 10 cm
so

3.25 dm = (10 × _____) cm
or

3.25 dm = _____ cm

18. 1 cm = 10 mm
so

51.2 cm = (_____ × 51.2) mm
or

51.2 cm = _____ mm

19. 1 mm = .1 cm
so

6 mm = (_____ × _____) cm
or

6 mm = _____ cm

20. 1 cm = .1 dm
so

1.25 cm = (_____ × _____) dm
or

1.25 cm = _____ dm

21. 17 dm = _____ cm   **24.** .5 cm = _____ mm   **27.** .13 cm = _____ dm

22. 2.4 cm = _____ dm   **25.** 62 mm = _____ cm   **28.** 7.5 mm = _____ cm

23. .45 dm = _____ cm   **26.** 42 cm = _____ mm   **29.** 1.2 dm = _____ cm

That'll be $.038, please.

Cents and Mills

You know that 1 dime = 10 cents.

But did you know that 1 cent = 10 mills?

A mill is a unit of money (but not an actual coin) used in accounting, especially taxes.

30. If one cent is shown as $.01, how would you show one

mill? _____

31. How are cents and mills similar to centimeters and

millimeters? _____

_____

Dekameter and Hectometer

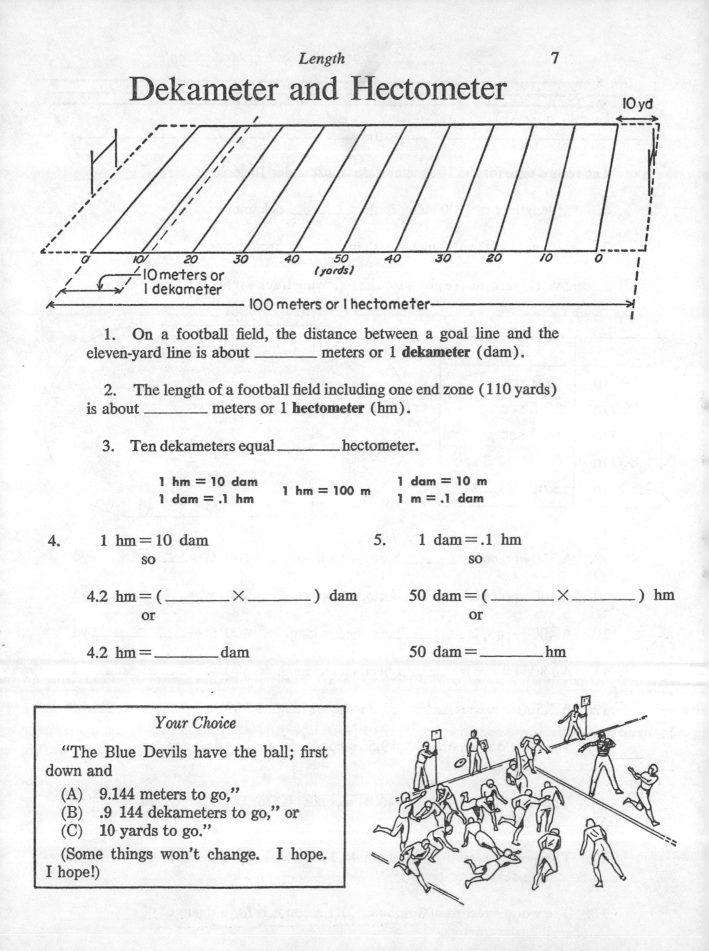

10 yd

10 meters or
1 dekameter

100 meters or 1 hectometer

(yards)

1.   On a football field, the distance between a goal line and the eleven-yard line is about _____ meters or 1 **dekameter** (dam).

2.   The length of a football field including one end zone (110 yards) is about _____ meters or 1 **hectometer** (hm).

3.   Ten dekameters equal _____ hectometer.

1 hm = 10 dam	1 hm = 100 m	1 dam = 10 m
1 dam = .1 hm		1 m = .1 dam

4.   1 hm = 10 dam
            so

4.2 hm = (_____ × _____) dam
        or

4.2 hm = _____ dam

5.   1 dam = .1 hm
            so

50 dam = (_____ × _____) hm
        or

50 dam = _____ hm

Your Choice

"The Blue Devils have the ball; first down and

(A)   9.144 meters to go,"
(B)   .9 144 dekameters to go," or
(C)   10 yards to go."

(Some things won't change.   I hope.
I hope!)

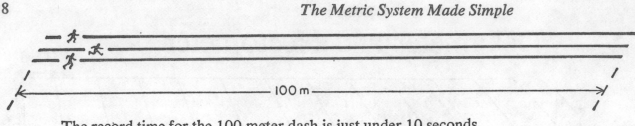

100 m

The record time for the 100-meter dash is just under 10 seconds.

6. The length of the 100-meter dash is _____ dekameters.

7. The length of the 100-meter dash is _____ hectometer.

Recent world records are given for the following track events.

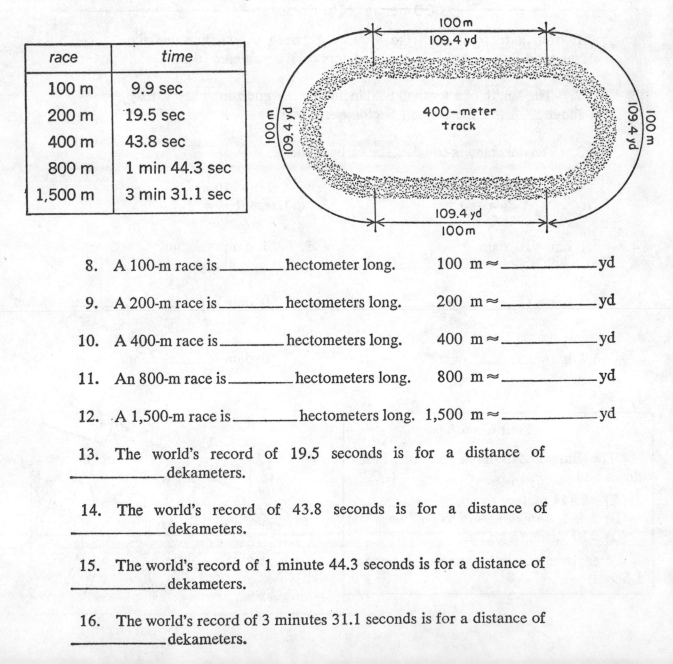

race	time
100 m	9.9 sec
200 m	19.5 sec
400 m	43.8 sec
800 m	1 min 44.3 sec
1,500 m	3 min 31.1 sec

8. A 100-m race is _____ hectometer long.   100 m ≈ _____ yd

9. A 200-m race is _____ hectometers long.   200 m ≈ _____ yd

10. A 400-m race is _____ hectometers long.   400 m ≈ _____ yd

11. An 800-m race is _____ hectometers long.   800 m ≈ _____ yd

12. A 1,500-m race is _____ hectometers long. 1,500 m ≈ _____ yd

13. The world's record of 19.5 seconds is for a distance of _____ dekameters.

14. The world's record of 43.8 seconds is for a distance of _____ dekameters.

15. The world's record of 1 minute 44.3 seconds is for a distance of _____ dekameters.

16. The world's record of 3 minutes 31.1 seconds is for a distance of _____ dekameters.

Kilometer

The 60-story Woolworth Building in New York is 792 feet high. Four such buildings, one on top of the other, would be about 1 **kilometer.**

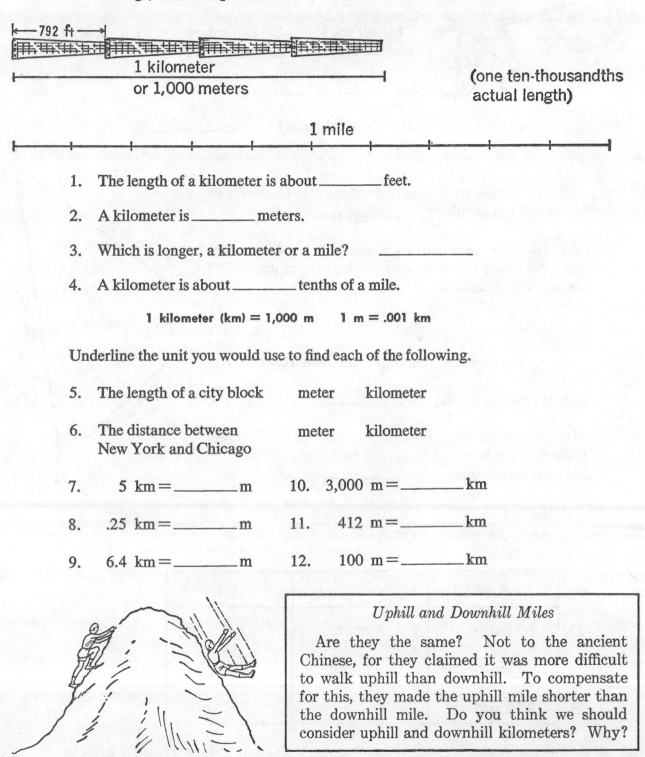

(one ten-thousandths actual length)

1. The length of a kilometer is about _____ feet.

2. A kilometer is _____ meters.

3. Which is longer, a kilometer or a mile?   _____

4. A kilometer is about _____ tenths of a mile.

1 kilometer (km) = 1,000 m   1 m = .001 km

Underline the unit you would use to find each of the following.

5. The length of a city block   meter   kilometer

6. The distance between   meter   kilometer
   New York and Chicago

7.   5 km = _____ m   10.   3,000 m = _____ km

8.   .25 km = _____ m   11.   412 m = _____ km

9.   6.4 km = _____ m   12.   100 m = _____ km

Uphill and Downhill Miles

Are they the same? Not to the ancient Chinese, for they claimed it was more difficult to walk uphill than downhill. To compensate for this, they made the uphill mile shorter than the downhill mile. Do you think we should consider uphill and downhill kilometers? Why?

Airline distances

13.   Mary flew from Chicago to Houston and then on to New York. Mike flew from Chicago to Miami and then on to New York. Who flew the greater distance?   _____

How much greater?   _____

14.   If you follow the paths shown on the map, what is the shortest route possible from San Francisco to New York?

_____

How far would it be?   _____

15.   Bob drove from A to B and then on to C. How many kilometers did he drive?   _____

16.   Bob made the trip in 4 hours. What was his average speed per hour? (total distance ÷ number of hours = average speed per hour)

_____

17.   Trudy traveled from B to C and then from C to A.

How far did she travel?   _____

18.   Trudy traveled an average speed of 65 km/h (read kilometers per hour). At that rate, how long did it take her to make the trip?   _____

Complete the following table.

	average speed	time traveled	total distance
19.	80 km/h	4 hours	
20.	88 km/h	5.5 hours	
21.	90 km/h		315 km

Prefixes and the Decimal System

The decimal 123.45 can be expressed as

$$123.45 = (1 \times 100) + (2 \times 10) + (3 \times 1) + (4 \times .1) + (5 \times .01)$$

$$= 1 \text{ hundred} + 2 \text{ tens} + 3 \text{ ones} + 4 \text{ tenths} + 5 \text{ hundredths}.$$

In a similar way, 123.45 meters can be expressed as

$$123.45 \text{ m} = (1 \times 100) \text{ m} + (2 \times 10) \text{ m} + (3 \times 1) \text{ m} + (4 \times .1) \text{ m} + (5 \times .01) \text{ m}$$

$$123.45 \text{ m} = 1 \text{ } hecto\text{meter} + 2 \text{ } deka\text{meters} + 3 \text{ meters} + 4 \text{ } deci\text{meters} + 5 \text{ } centi\text{meters}$$

Notice the similarity between these expanded forms. The units of length are related to the meter, the basic unit, in the same way that the place values (hundreds, tens, tenths, hundredths) are related to the ones (or units) place of the decimal system. These relationships are clearly shown in the following chart.

place values	thousands (th)	hundreds (h)	tens (t)	ones	tenths (ts)	hundredths (hs)	thousandths (ths)
meaning	1,000	100	10	1	.1	.01	.001
metric prefixes	kilo	hecto	deka	basic unit	deci	centi	milli

Prefixes indicating multiples greater than one are derived from Greek words. Prefixes indicating multiples less than one are derived from Latin words. Tell whether each of the following was derived from a Latin word or from a Greek word.

1. centi _____
2. deka _____
3. kilo _____

4. milli _____
5. hecto _____
6. deci _____

Certain money values correspond directly to metric prefixes. Write the appropriate money value after each prefix below. Choose from these values: $1,000, $100, $10, $.10, $.01, and $.001.

7. kilo _____   10. centi _____

8. milli _____   11. hecto _____

9. deci _____   12. deka _____

km   hm   dam   m   dm   cm   mm

The metric system is based on decimal numeration. Therefore, each unit is 10 times greater than the next smaller unit and .1 as great as the next larger unit.

13.   4   th = 40   h = 400   t   = _____ ones = _____ ts

14.   4   km = _____ hm = _____ dam = _____ m = _____ dm

15.   400 t = 4,000 ones = _____ ts = _____ hs = _____ ths

16.   400 dam = _____ m = _____ dm = _____ cm = _____ mm

17. _____ h = _____ t = 2 ones = _____ ts = _____ hs

18. _____ hm = _____ dam = 2 m = _____ dm = _____ cm

19.   1 m = _____ dm   27. .4 dam = _____ hm

20.   1 hm = _____ dam   28.   1 m = _____ cm

21.   1 cm = _____ mm   29.   1 dam = _____ dm

22.   1 hm = _____ km   30.   1 dm = _____ mm

23.   1 m = _____ dam   31.   1 m = _____ hm

24.   1 cm = _____ dm   32.   1 cm = _____ m

25.   .13 hm = _____ dam   33.   1 dam = _____ km

26.   .02 dm = _____ m   34.   .2 dm = _____ dam

35. .14 dm = _____ mm

36. .35 hm = _____ m

37. 1 km = _____ m

38. 1 dm = _____ hm

39. 1 dam = _____ mm

40. 2 km = _____ mm

41. 2 cm = _____ hm

42. 2 mm = _____ m

43. 3 m = _____ km

44. 3 hm = _____ dm

45. 3 cm = _____ km

Estimating

Guess how long this line segment is in centimeters. Record your *estimate*.

1. _____ cm

Now measure the line segment. Record its *actual* length.

2. _____ cm

Estimate the length of each line segment. Then use a centimeter ruler to find the actual length.

	estimate	*actual*
3.	_____ cm	_____ cm
4.	_____ cm	_____ cm
5.	_____ cm	_____ cm

Estimate the length of each line segment in millimeters. Then use a millimeter ruler to find the actual length.

	estimate	*actual*
6.	_____ mm	_____ mm
7.	_____ mm	_____ mm
8.	_____ mm	_____ mm

Use the millimeter ruler to draw line segments of the following lengths.

9. 16 cm

10. 97 mm

11. 5 cm 8 mm

12. 13 cm 4 mm

Estimate the length of each object in centimeters.
Then find the length of each object to the nearest centimeter.

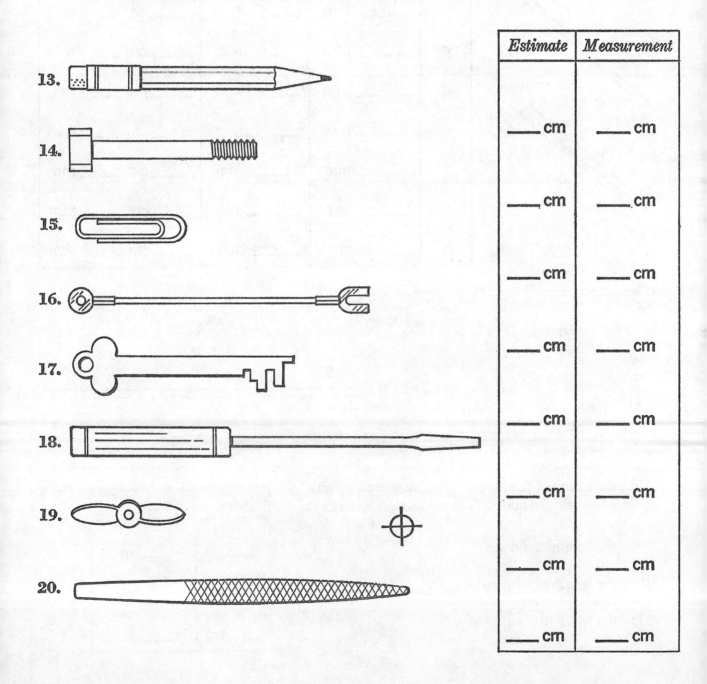

Estimate	Measurement
____ cm	____ cm
____ cm	____ cm
____ cm	____ cm
____ cm	____ cm
____ cm	____ cm
____ cm	____ cm
____ cm	____ cm
____ cm	____ cm

Estimate the length of each object in millimeters.
Then find the length to the nearest millimeter. Record
each measurement in two ways as shown in the table.

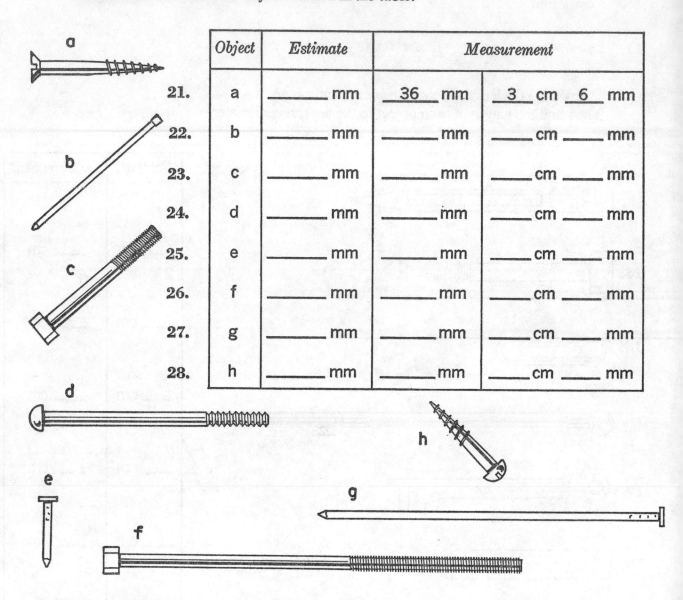

Object	Estimate	Measurement		
21. a	_____ mm	__36__ mm	__3__ cm	__6__ mm
22. b	_____ mm	_____ mm	_____ cm	_____ mm
23. c	_____ mm	_____ mm	_____ cm	_____ mm
24. d	_____ mm	_____ mm	_____ cm	_____ mm
25. e	_____ mm	_____ mm	_____ cm	_____ mm
26. f	_____ mm	_____ mm	_____ cm	_____ mm
27. g	_____ mm	_____ mm	_____ cm	_____ mm
28. h	_____ mm	_____ mm	_____ cm	_____ mm

Estimate each of the following to the nearest meter.
Use a meter stick to find each to the nearest meter.

	Estimate	Measurement
29. the height of a door	____ m	____ m
30. the length of your room	____ m	____ m
31. the width of your room	____ m	____ m

Adding Metric Measures

Tanya found the distance across a quarter.
José found the distance across a penny. They
laid the coins beside each other as shown. How
can they find the distance across both coins?

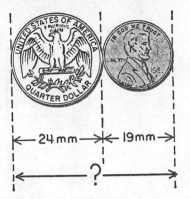

One way is to measure across them.

1. The distance is _____ mm.

Another way is to add the two measures.

$$\begin{array}{r} 24 \text{ mm} \\ +19 \text{ mm} \\ \hline 43 \text{ mm} \end{array}$$

Tanya and José could have given the measurements as follows.

2. 24 mm = _____ cm _____ mm

3. 19 mm = _____ cm _____ mm

Then the measures could be added as follows.

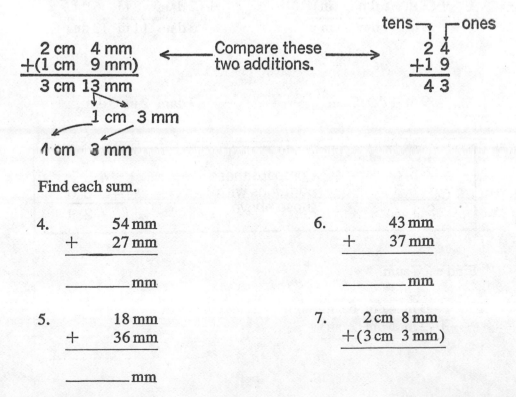

Find each sum.

4. $\begin{array}{r} 54 \text{ mm} \\ + \quad 27 \text{ mm} \\ \hline \\ _____ \text{ mm} \end{array}$

6. $\begin{array}{r} 43 \text{ mm} \\ + \quad 37 \text{ mm} \\ \hline \\ _____ \text{ mm} \end{array}$

5. $\begin{array}{r} 18 \text{ mm} \\ + \quad 36 \text{ mm} \\ \hline \\ _____ \text{ mm} \end{array}$

7. $\begin{array}{r} 2 \text{ cm} \quad 8 \text{ mm} \\ +(3 \text{ cm} \quad 3 \text{ mm}) \\ \hline \end{array}$

8.　　7 cm　5 mm
　　+(1 cm　7 mm)
　　――――――――

9.　　4 cm　9 mm
　　+(2 cm　8 mm)
　　――――――――

10.　　　　16 cm
　　+　　47 cm
　　――――――――

　―――――― cm　or ―― dm ―――――― cm

11.　　　　38 cm
　　+　　15 cm
　　――――――――

　―――――― cm　or ―― dm ―――――― cm

12.　　　　54 cm
　　+　　39 cm
　　――――――――

　―――――― cm　or ―― dm ―――――― cm

Study how these measures are added.

7 m　5 dm　4 cm	4 dam　3 m　7 dm
+(1 m　6 dm　3 cm)	+(2 dam　8 m　6 dm)
8 m　11 dm　7 cm	6 dam　11 m　13 dm

　　　　　　or　　　　　　　　　　　　　　or

　9 m　1 dm　7 cm　　　　　　　7 dam　2 m　3 dm

```
  7 5 4          Compare these          4 3 7
 +1 6 3          additions with        +2 8 6
 ─────           those above.          ─────
  9 1 7                                 7 2 3
```

Find each sum.

13.　　3 m　4 dm　6 cm
　　+(2 m　5 dm　2 cm)

14.   4 m   1 dm   2 cm
    +(3 m   4 dm   9 cm)

20.   6 dam   7 m   6 dm
    +(1 dam   1 m   5 dm)

15.   1 m   6 dm   3 cm
    +(6 m   4 dm   8 cm)

21.   3 dam   8 m   4 dm
    +(5 dam   3 m   9 dm)

16.   6 dm   3 cm   4 mm
    +(2 dm   5 cm   7 mm)

22.   1 hm   4 dam   6 m
    +(3 hm   8 dam   4 m)

17.   4 dm   6 cm   5 mm
    +(2 dm   4 cm   8 mm)

23.   2 hm   9 dam   4 m
    +(5 hm   4 dam   3 m)

18.   5 dm   9 cm   7 mm
    +(1 dm   3 cm   6 mm)

24.   2 km   6 hm   7 dam
    +(4 km   5 hm   8 dam)

19.   4 dam   3 m   8 dm
    +(2 dam   6 m   1 dm)

Subtracting Metric Measures

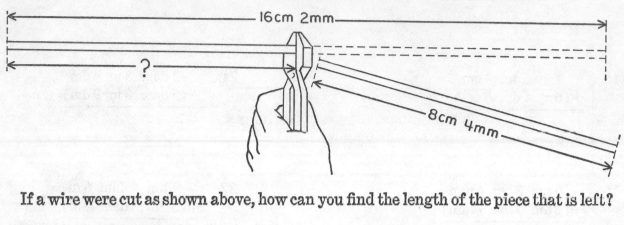

If a wire were cut as shown above, how can you find the length of the piece that is left?

One way is to measure it.

1. Its length is _____ cm _____ mm.

Another way is to subtract the measures.

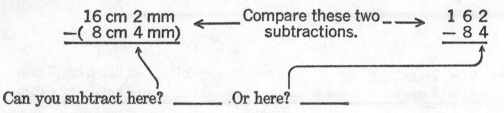

$$\begin{array}{r} 16 \text{ cm } 2 \text{ mm} \\ -(\ 8 \text{ cm } 4 \text{ mm}) \\ \hline \end{array}$$ Compare these two subtractions. $$\begin{array}{r} 1\ 6\ 2 \\ -\ 8\ 4 \\ \hline \end{array}$$

2. Can you subtract here? _____ .Or here? _____

3. Rename 16 cm 2 mm as 15 cm _____ mm.

Then subtract the measures.

$$\begin{array}{r} 15 \text{ cm } 12 \text{ mm} \\ -(\ 8 \text{ cm }\ \ 4 \text{ mm}) \\ \hline 7 \text{ cm }\ \ 8 \text{ mm} \end{array}$$ Compare these two subtractions. $$\begin{array}{r} \overset{5\ \ 12}{1\ \cancel{6}\ \cancel{2}} \\ -\ 8\ 4 \\ \hline 7\ 8 \end{array}$$

Find each difference.

4.   32 cm
      −17 cm

5.   44 mm
      −25 mm

6.   66 m
      −47 m

7.   9 cm 3 mm
      −(4 cm 2 mm)

8.   8 cm 4 mm
      −(5 cm 7 mm)

9.   7 cm 3 mm
      −(6 cm 5 mm)

10.   5 dm 1 cm
      −(1 dm 4 cm)

11.   4 m 8 dm
      −(2 m 9 dm)

Study how these measures are subtracted.

$$\begin{array}{c} {}^{7}\!\!\!\!\!\!\!\!\!\!\!\!\ {}^{12}\\ \cancel{8}\,\text{dm}\ \cancel{2}\,\text{cm}\ 6\,\text{mm}\\ -(3\,\text{dm}\ 7\,\text{cm}\ 4\,\text{mm})\\ \hline 4\,\text{dm}\ 5\,\text{cm}\ 2\,\text{mm} \end{array}$$

$$\begin{array}{c} {}^{6}\ {}^{13}\ {}^{11}\\ \cancel{7}\,\text{m}\ \cancel{4}\,\text{dm}\ \cancel{1}\,\text{cm}\\ -(5\,\text{m}\ 7\,\text{dm}\ 3\,\text{cm})\\ \hline 1\,\text{m}\ 6\,\text{dm}\ 8\,\text{cm} \end{array}$$

$$\begin{array}{c} {}^{7}\ {}^{12}\\ 8\ \cancel{2}\ 6\\ -3\ 7\ 4\\ \hline 4\ 5\ 2 \end{array}$$

Compare these subtractions with those above.

$$\begin{array}{c} {}^{6}\ {}^{13}\ {}^{11}\\ \cancel{7}\ \cancel{4}\ \cancel{1}\\ -5\ 7\ 3\\ \hline 1\ 6\ 8 \end{array}$$

Find each difference.

12. 17 m 6 dm
−(9 m 8 dm)

13. 7 dm 8 cm 5 mm
−(3 dm 6 cm 2 mm)

14. 6 dm 7 cm 3 mm
−(2 dm 4 cm 8 mm)

15. 4 dm 3 cm 9 mm
−(1 dm 6 cm 7 mm)

16. 8 dm 1 cm 2 mm
−(3 dm 5 cm 4 mm)

17. 9 m 3 dm 4 cm
−(4 m 6 dm 7 cm)

18. 5 m 8 dm 3 cm
−(2 m 7 dm 6 cm)

19. 6 dam 3 m 9 dm
−(2 dam 5 m 7 dm)

20. 8 dam 4 m 7 dm
−(6 dam 8 m 9 dm)

21. 7 hm 5 dam 6 m
−(4 hm 3 dam 8 m)

22. 4 hm 2 dam 1 m
−(1 hm 6 dam 3 m)

23. 5 km 8 hm 4 dam
− (2 km 9 hm 2 dam)

24. 9 km 3 hm 5 dam
−(6 km 7 hm 8 dam)

Problems About Length

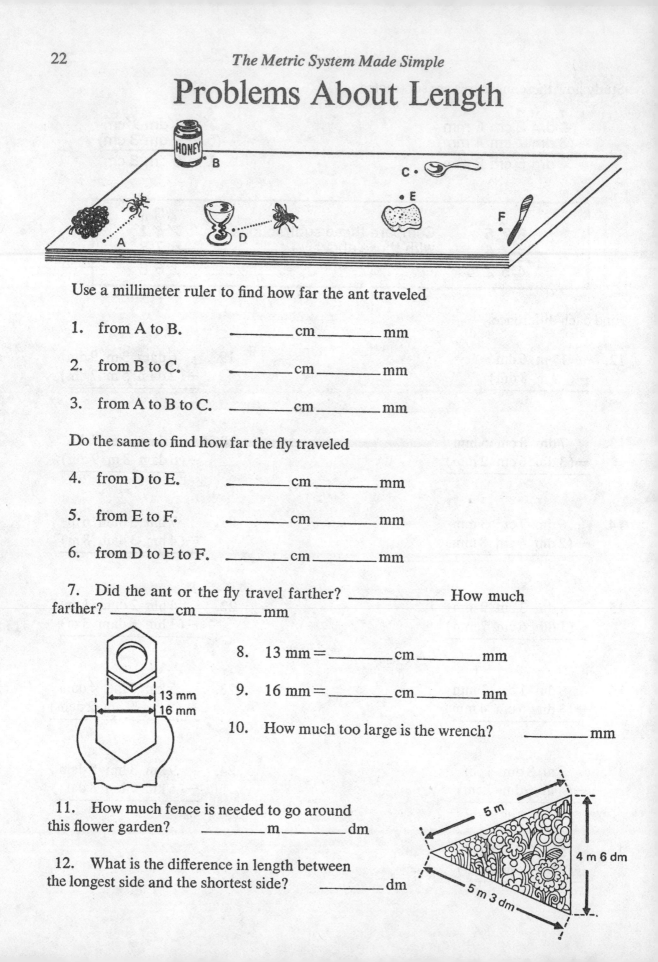

Use a millimeter ruler to find how far the ant traveled

1. from A to B.   ——————— cm ———————mm

2. from B to C.   ——————— cm ———————mm

3. from A to B to C.   ——————— cm ———————mm

Do the same to find how far the fly traveled

4. from D to E.   ——————— cm ———————mm

5. from E to F.   ——————— cm ———————mm

6. from D to E to F.   ——————— cm ———————mm

7. Did the ant or the fly travel farther? ——————— How much farther? ———————cm ———————mm

8. 13 mm = ———————cm ———————mm

9. 16 mm = ———————cm ———————mm

10. How much too large is the wrench?   ———————mm

11. How much fence is needed to go around this flower garden? ———————m ———————dm

12. What is the difference in length between the longest side and the shortest side? ———————dm

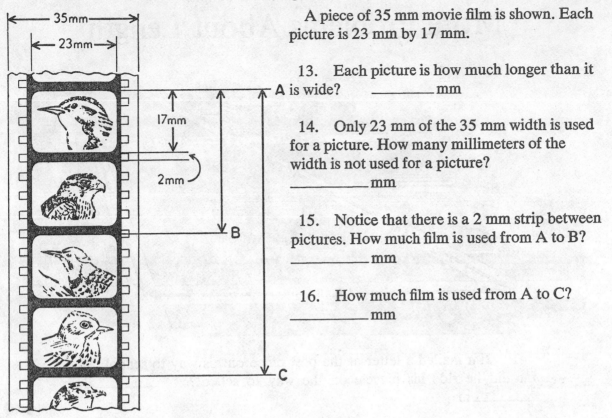

A piece of 35 mm movie film is shown. Each picture is 23 mm by 17 mm.

13. Each picture is how much longer than it is wide? _____ mm

14. Only 23 mm of the 35 mm width is used for a picture. How many millimeters of the width is not used for a picture?
_____ mm

15. Notice that there is a 2 mm strip between pictures. How much film is used from A to B?
_____ mm

16. How much film is used from A to C?
_____ mm

One model of a Mercedes car has the measurements shown below.

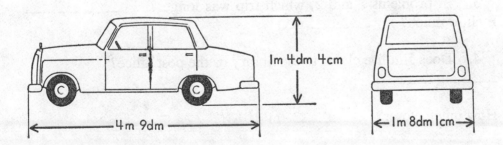

17. This car is _____ m _____ dm _____ cm longer than it is wide.

18. This car is _____ dm _____ cm wider than it is high.

19. How long would two of these cars be when parked bumper to bumper? _____ m _____ dm

20. A highway sign was twice as tall as this car. How tall was the sign? _____ m _____ dm _____ cm

More Problems About Length

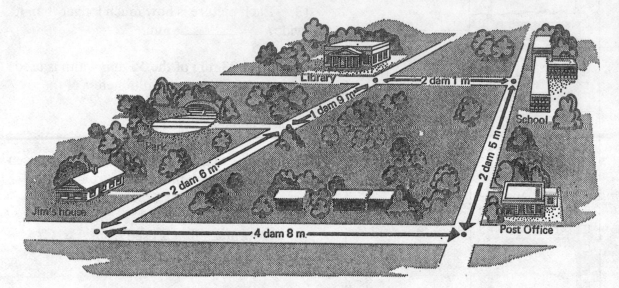

1. Jim mailed a letter at the post office on his way to school. How far did he ride his bicycle on the way to school? _____ dam _____ m

2. He returned a book to the library on his way home from school. How far did he ride on his way home? _____ dam _____ m

3. In problems *1* and *2*, which trip was longer? _____ By how much? _____ m

4. Does Jim live closer to the library or the post office? _____

How much closer? _____ m

Shown below are the odometer readings in kilometers on Mr. Chenault's car before and after a trip.

Before

After

5. How long was the trip? _____ km

6. Mr. Chenault made the trip in 9 hours. What was his average speed? _____ km per hour

7. After driving 196 kilometers, he got on the tollway for the rest of the trip. How far did he drive on the tollway? _____ km

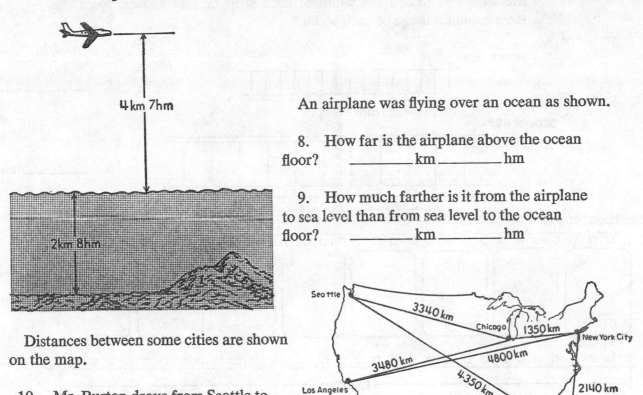

An airplane was flying over an ocean as shown.

8. How far is the airplane above the ocean floor? _____ km _____ hm

9. How much farther is it from the airplane to sea level than from sea level to the ocean floor? _____ km _____ hm

Distances between some cities are shown on the map.

10. Mr. Burton drove from Seattle to Chicago and then on to New York City.

How far did he drive? _____ km

11. Mary Ludens has to drive from Los Angeles to New York City. How much shorter is it to go direct than to go through Chicago? _____ km

12. How much farther is it from New York City to Miami than it is from New York City to Chicago? _____ km

13. The Jensens live in Seattle. On their vacation they went to Chicago, New York City, Miami, and then back to Seattle. How far did they go? _____ km

14. A Chicago team went to New York City for a game on Monday. They had games in Miami on Tuesday and Wednesday. On Thursday they played again in New York City. They had a home game on Friday. How far did the team travel on this trip? _____ km

The size of movie film is given by naming its width in millimeters. Measure and record the width of each strip of film shown. Note the most common usage of each width.

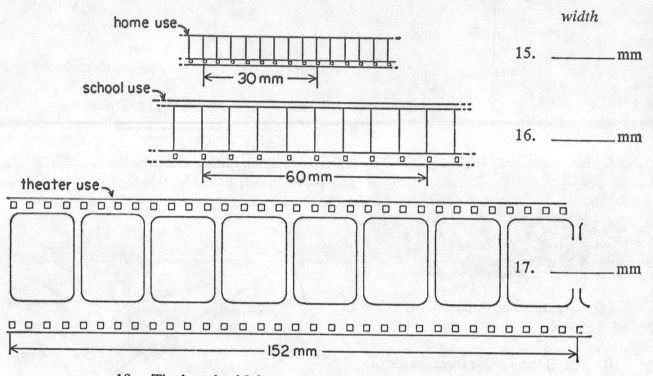

width

15. _____ mm

16. _____ mm

17. _____ mm

18. The length of 8 frames of an 8 mm film is _____ mm.

19. The length of 24 frames of an 8 mm film is _____ mm.

Movie projectors for the film sizes pictured above are designed to show 24 frames every second.

20. How long would an 8 mm strip of film have to be if it were to run exactly 1 second? _____ mm

21. How long would an 8 mm strip of film have to be if it were to run exactly 1 minute? _____ mm

Complete the table for each film size given.

	film size	length of 8 frames (in mm)	length of 24 frames (in mm)	length of film needed for 1 second	length of film needed for 1 minute
22.	16 mm	_____ mm	_____ mm	_____ mm	_____ mm
23.	35 mm	_____ mm	_____ mm	_____ mm	_____ mm

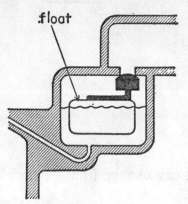

.float

24. The float level of this carburetor should be adjusted to 18.5 mm. To operate effectively, the adjustment should not vary more than 1.0 mm. What is the highest level at which the float could be adjusted? _____

25. In problem 24, what is the lowest level at which the float could be adjusted? _____

26. B indicates the width of the *valve seat*. The width of the intake valve seat should be 1.25 mm, with a possible variance of ± 0.15 mm. (± means *plus or minus*.) What are the maximum and minimum seat widths possible to meet these requirements?

maximum: _____   minimum: _____

27. The width of the exhaust valve seat should be 1.55 mm with a possible variance of ± 0.1 mm. What are the maximum and minimum seat widths possible to meet these requirements?

maximum: _____   minimum: _____

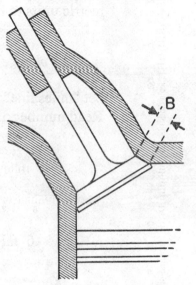

B

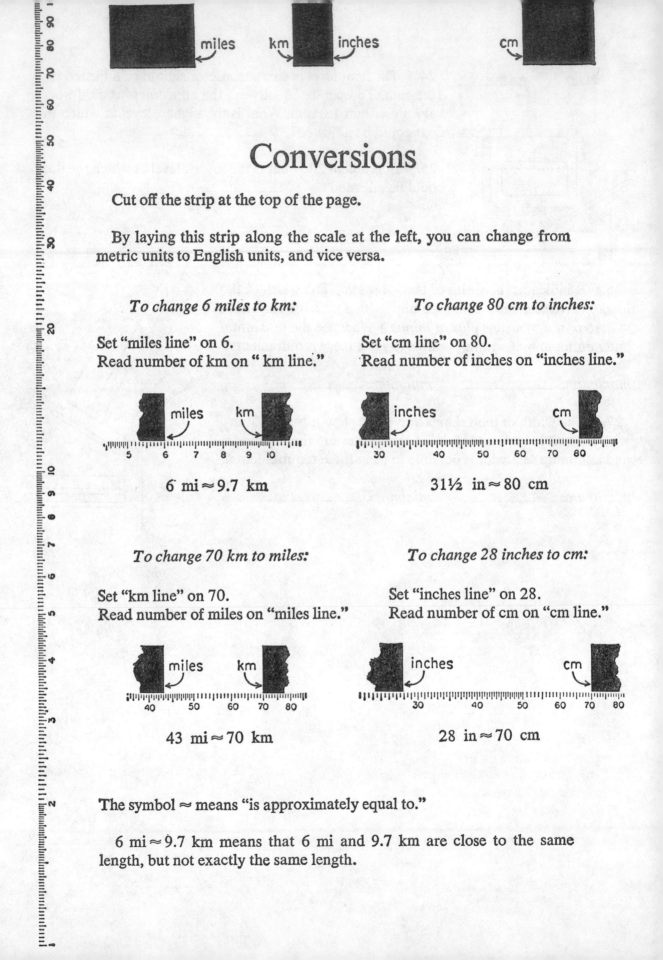

Conversions

Cut off the strip at the top of the page.

By laying this strip along the scale at the left, you can change from metric units to English units, and vice versa.

To change 6 miles to km:

Set "miles line" on 6.
Read number of km on " km line."

$$6 \text{ mi} \approx 9.7 \text{ km}$$

To change 80 cm to inches:

Set "cm line" on 80.
Read number of inches on "inches line."

$$31\tfrac{1}{2} \text{ in} \approx 80 \text{ cm}$$

To change 70 km to miles:

Set "km line" on 70.
Read number of miles on "miles line."

$$43 \text{ mi} \approx 70 \text{ km}$$

To change 28 inches to cm:

Set "inches line" on 28.
Read number of cm on "cm line."

$$28 \text{ in} \approx 70 \text{ cm}$$

The symbol $\approx$ means "is approximately equal to."

$6 \text{ mi} \approx 9.7 \text{ km}$ means that 6 mi and 9.7 km are close to the same length, but not exactly the same length.

1. 2 mi ≈ _____ km

2. 30 mi ≈ _____ km

3. 19 mi ≈ _____ km

4. _____ mi ≈ 8 km

5. _____ mi ≈ 50 km

6. _____ mi ≈ 13 km

7. _____ in ≈ 5 cm

8. _____ in ≈ 20 cm

9. _____ in ≈ 9 cm

10. 7 in ≈ _____ cm

11. 11 in ≈ _____ cm

12. 35 in ≈ _____ cm

The following is an alternate method for making conversions.

By setting the circular slide-rule computer as shown, you can convert miles to kilometers and vice versa.

Find 10 on the *miles* (outer) *scale*. The corresponding numeral on the *kilometers* (inner) *scale* is 16. Thus, you know 10 mi ≈ 16 km.

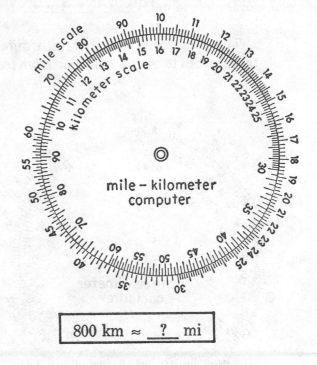

If **10** is considered as		Then **16** is considered as	
1 mi	≈	**16** km
1 mi	≈	**1.6** km
10 mi	≈	**16** km
100 mi	≈	**160** km

3 mi ≈ __?__ km

800 km ≈ __?__ mi

Find 30 on the miles scale.

13. The corresponding numeral on the kilometer scale is _____ .

14. 30 mi ≈ 48 km
    so

    3.0 mi ≈ _____ km

Find 80 on the kilometers scale.

15. The corresponding numeral on the miles scale is _____ .

16. 80 km ≈ 50 mi
    so

    800 km ≈ _____ mi

Use the miles-kilometers computer to help you complete the following.

17.   15 mi ≈ _____ km   23.   32 km ≈ _____ mi

18.   150 mi ≈ _____ km   24.   1,600 km ≈ _____ mi

19.   220 mi ≈ _____ km   25.   27 km ≈ _____ mi

20.   75 mi ≈ _____ km   26.   400 km ≈ _____ mi

21.   2,000 mi ≈ _____ km   27.   510 km ≈ _____ mi

22.   44 mi ≈ _____ km   28.   75 km ≈ _____ mi

The same computer, but with different settings, is used to convert from inches to metric units and from feet to metric units.

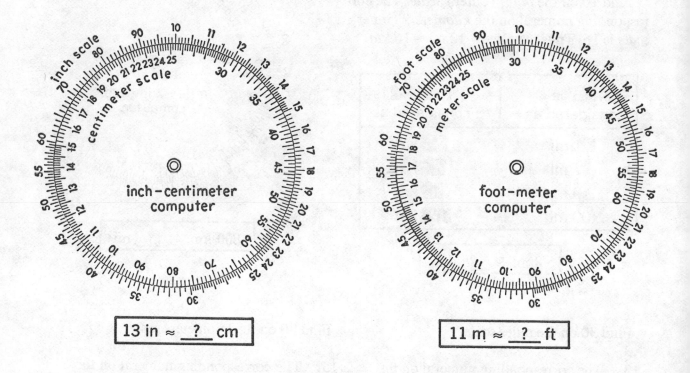

| 13 in ≈ _?_ cm | 11 m ≈ _?_ ft |

Find 13 on the inch scale.   Find 11 on the metric scale.

29.   The corresponding numeral on the metric scale is _____ .

31.   The corresponding numeral on the foot scale is _____ .

30.   13 in ≈ _____ cm

32.   11 m ≈ _____ ft

> Placement of the decimal point determines if you convert to centimeters, decimeters, meters, and so on.

33. Since 13 in ≈ __33__ cm,   34. Since 11 m ≈ __36__ ft,

    then 13 in ≈ _____ m.   then 1,100 cm ≈ _____ ft.

Use the appropriate computer to help you complete the following.

35.   31 in ≈ _____ cm = _____ m

36.   .9 m = _____ cm ≈ _____ in

37.   620 in ≈ _____ cm = _____ m

38   8.5 m = _____ cm ≈ _____ in

39   160 in ≈ _____ cm = _____ m

40. 1,400 cm = _____ m ≈ _____ ft

41.   410 ft ≈ _____ m = _____ cm

42   49 ft ≈ _____ m = _____ cm

43. 1,600 cm = _____ m ≈ _____ ft

44.   95 ft ≈ _____ m = _____ cm

Check Your Progress

Measure each line segment to the nearest centimeter.

1. _____ _____ cm

2. _____ _____ cm

3. _____ _____ cm

Measure each line segment to the nearest millimeter.

4. _____ _____ mm

5. _____ _____ mm

6. _____ _____ mm

Measure each line segment to the nearest centimeter.
Then measure it to the nearest millimeter.

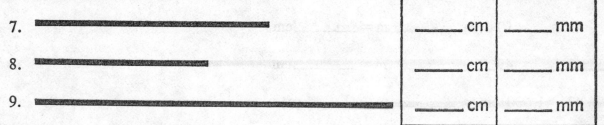

7. _____   _____ cm   _____ mm

8. _____   _____ cm   _____ mm

9. _____   _____ cm   _____ mm

Complete the following chart.

	metric prefix	meaning	decimal place value
	milli	.001	thousandths
10.	centi		
11.		.1	
12.			tens
13.	hecto		
14.		1,000	

Complete the following.

15. 1 m = _____ dm

16. 1 m = _____ cm

17. 1 m = _____ mm

18. 1 dm = _____ cm

19. 1 cm = _____ mm

20. 1 km = _____ hm

21. 1 km = _____ dam

22. 1 km = _____ m

23. 1 hm = _____ dam

24. 1 dam = _____ m

25. 4 m = _____ dm

26. 3 m = _____ cm

27. .5 km = _____ m

28. 14 m = _____ km

29. 23.5 mm = _____ m

30. 14.2 mm = _____ cm

Find each sum or difference.

31.   3 m 6 dm 4 cm
    +(2 m 1 dm 7 cm)
    _____

32.   5 dm 7 cm 6 mm
    +(1 dm 5 cm 8 mm)
    _____

33.   8 m 3 dm 4 cm
    −(3 m 7 dm 1 cm)
    _____

34.   6 dm 2 cm 3 mm
    −(4 dm 3 cm 9 mm)
    _____

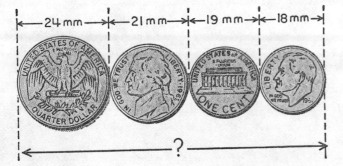

35. How far is it across all four coins?

_____ cm _____ mm

36. Omar jumped 6 m 5 dm 6 cm. José jumped 7 m 1 dm 2 cm. How much farther did José jump?

_____ dm _____ cm

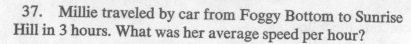

Foggy Bottom

Rainy Day

264 km

Sunset Lake

314 km

Sunrise Hill

192 km

Cloudy Gap

37.   Millie traveled by car from Foggy Bottom to Sunrise Hill in 3 hours. What was her average speed per hour?

_____

38.   Millie continued her trip from Sunrise Hill to Cloudy Gap by train and then on to Rainy Day by bus. How far did she travel by car, train, and bus?   _____

39.   Did Millie travel farther by train or by bus?   _____

How much farther?   _____

40.   When Millie continued from Rainy Day to Foggy Bottom to complete her trip, she had traveled a total of 923 kilometers. How far is it from Rainy Day to Foggy Bottom?

_____

41.   Is the length of each edge of this negative the same?

_____

42.   What is the length of each edge?   _____

Perimeter and Area

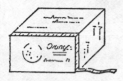

How far
around ?

Perimeter

How much
surface?

The **perimeter** of a figure is the distance around it.

The perimeter of ABCD is 6 + 3 + 6 + 3 or 18 cm.

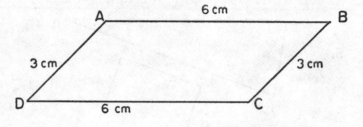

Estimate the perimeter of each figure. Then use a centimeter ruler to find the actual perimeter.

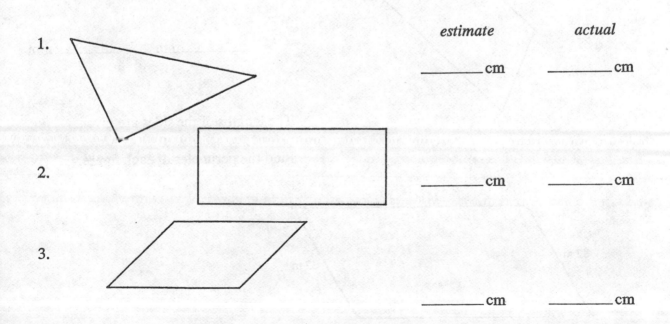

	estimate	actual
1.	_____ cm	_____ cm
2.	_____ cm	_____ cm
3.	_____ cm	_____ cm

Estimate the perimeter of each figure in millimeters. Then use a millimeter ruler to find the actual perimeter.

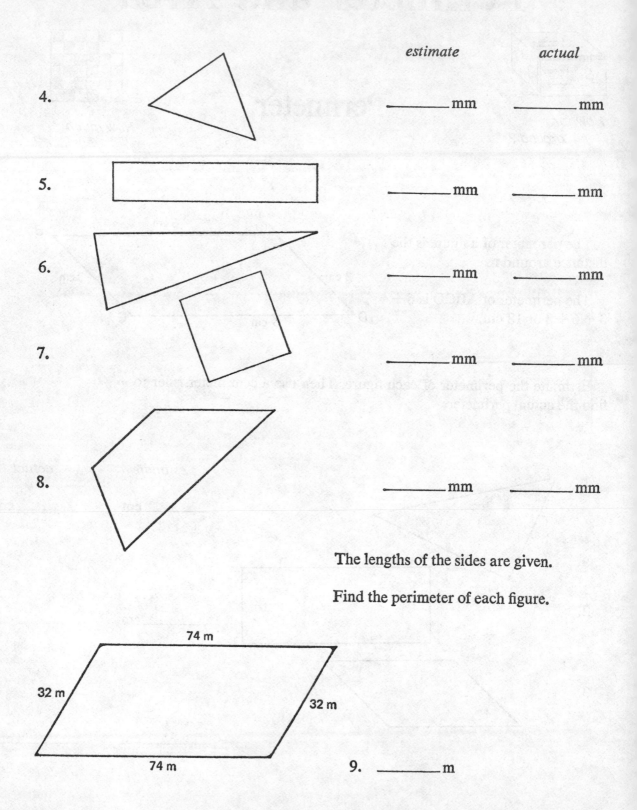

estimate   actual

4.   ———————— mm   ———————— mm

5.   ———————— mm   ———————— mm

6.   ———————— mm   ———————— mm

7.   ———————— mm   ———————— mm

8.   ———————— mm   ———————— mm

The lengths of the sides are given.

Find the perimeter of each figure.

74 m

32 m

32 m

74 m   9. ———————— m

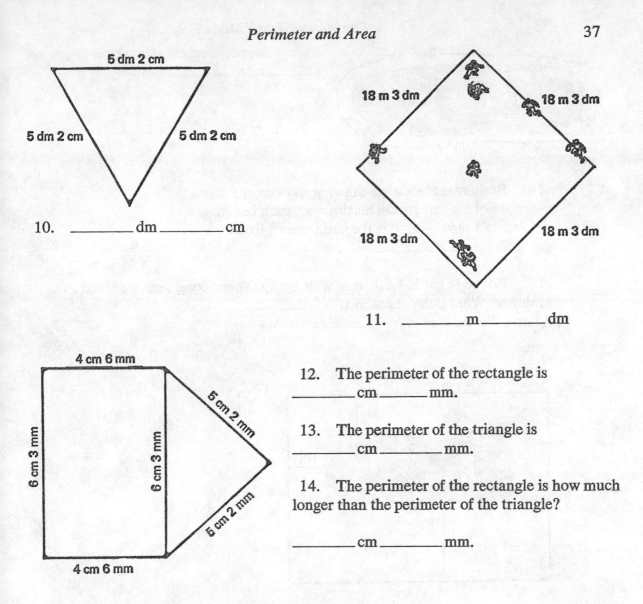

5 dm 2 cm

5 dm 2 cm   **5 dm 2 cm**

10. _____ dm _____ cm

18 m 3 dm   **18 m 3 dm**

18 m 3 dm   **18 m 3 dm**

11. _____ m _____ dm

4 cm 6 mm

6 cm 3 mm   **6 cm 3 mm**   **5 cm 2 mm**

5 cm 2 mm

4 cm 6 mm

12.   The perimeter of the rectangle is _____ cm _____ mm.

13.   The perimeter of the triangle is _____ cm _____ mm.

14.   The perimeter of the rectangle is how much longer than the perimeter of the triangle?

_____ cm _____ mm.

The perimeter is given. Find the missing length.

Perimeter is 43 cm.

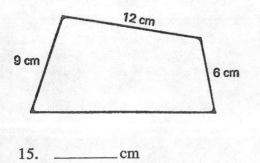

12 cm

9 cm   **6 cm**

15. _____ cm

Perimeter is 14 dm 3 cm.

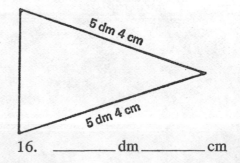

5 dm 4 cm

5 dm 4 cm

16. _____ dm _____ cm

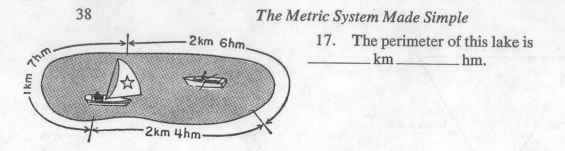

17. The perimeter of this lake is
_____ km _____ hm.

18. Rodney made a scale drawing showing the dimensions of his garage. On his drawing, each centimeter represents 1 meter. What is the perimeter of Rodney's garage? _____

19. Rodney's car is 1,800 mm wide by 5,000 mm long and is parked as shown. What is the distance *x*? _____
What is the distance *y*? _____

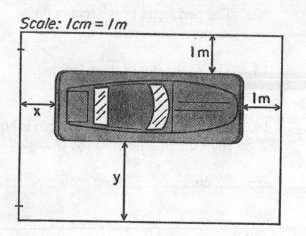

Scale: 1cm = 1m

Area

Square Units

Square units are used to measure area.

The following square units are actual size.

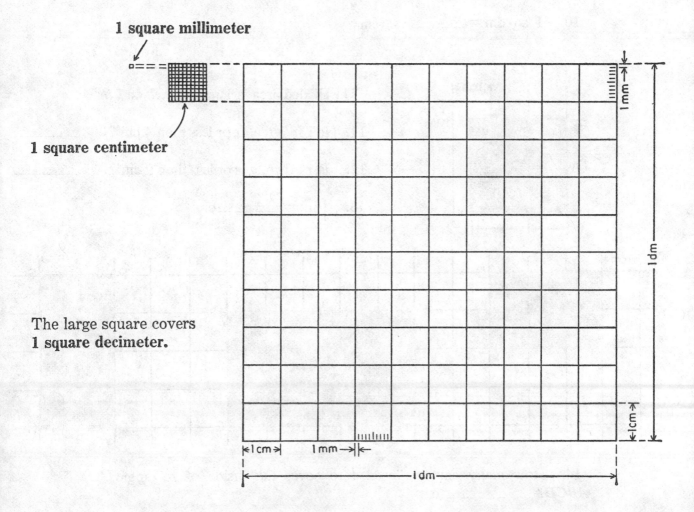

1 square millimeter

1 square centimeter

The large square covers
1 square decimeter.

You can abbreviate *square centimeter* as **cm²**.

1. You can write *square millimeter* as _____.

2. You can write *square decimeter* as _____.

3. 1 cm = _____ mm

4. 1 cm² = _____ mm²

5. 1 dm = _____ cm

6. 1 dm² = _____ cm²

You can also write *square centimeter* as **sq cm.**

7. You can write *square millimeter* as _____.

8. You can write *square decimeter* as _____.

9. 1 sq cm = _____ sq mm

10. 1 sq dm = _____ sq cm

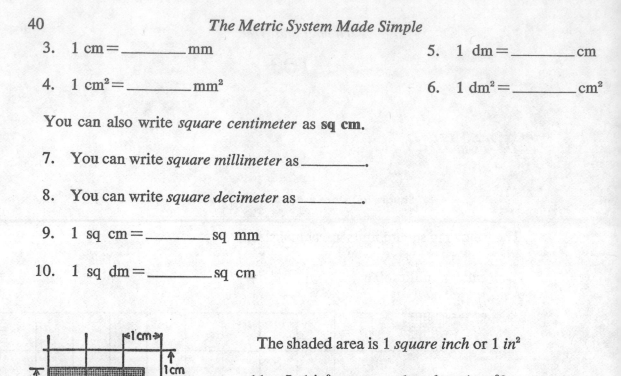

The shaded area is 1 *square inch* or 1 *in²*

11. Is 1 in² greater or less than 4 cm²? _____

12. Is 1 in² greater or less than 9 cm²? _____

13. 1 in² = _____ cm²

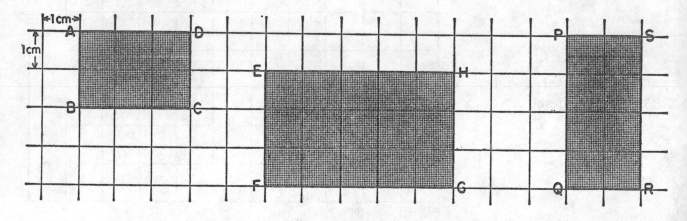

14. How many cm² are needed to cover the inside of rectangle *ABCD?* _____ cm²

We say that the **area** of rectangle *ABCD* is 6 cm².
(length × width = area)

15. What is the area of rectangle *EFGH?* _____ cm²

16. What is the area of rectangle *PQRS?* _____ cm²

Measure and record the length and width of each rectangle in the units indicated. Then find the area of each figure.

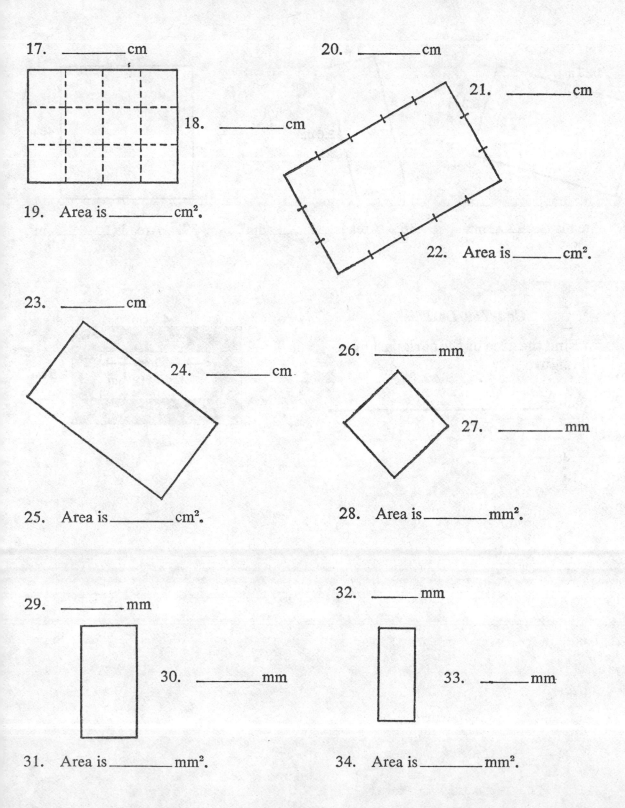

17. _____ cm

18. _____ cm

19. Area is _____ cm².

20. _____ cm

21. _____ cm

22. Area is _____ cm².

23. _____ cm

24. _____ cm

25. Area is _____ cm².

26. _____ mm

27. _____ mm

28. Area is _____ mm².

29. _____ mm

30. _____ mm

31. Area is _____ mm².

32. _____ mm

33. _____ mm

34. Area is _____ mm².

Find the area of each rectangular figure.

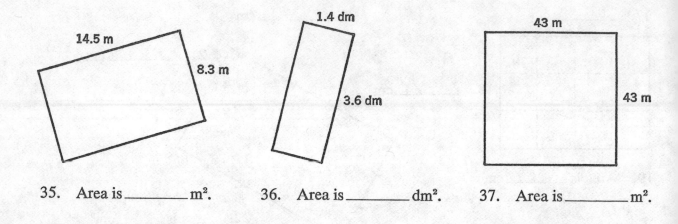

35. Area is _____ m².   36. Area is _____ dm².   37. Area is _____ m².

Can You Do This?

Find the area of the dark part of
the figure.

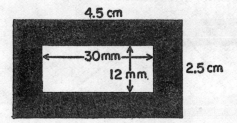

38. _____

Area of a Rectangle

Answer the following for this rectangle.

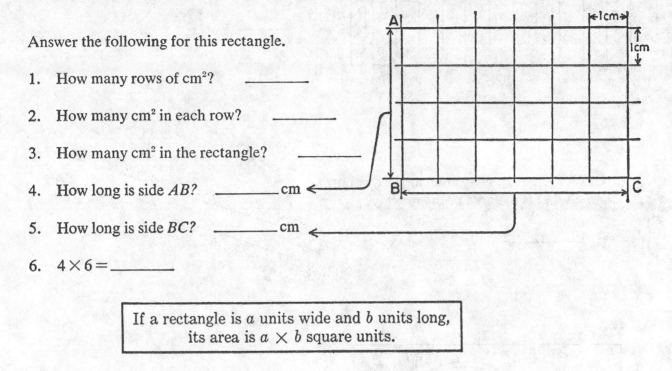

1. How many rows of cm²? _____

2. How many cm² in each row? _____

3. How many cm² in the rectangle? _____

4. How long is side *AB?* _____ cm

5. How long is side *BC?* _____ cm

6. $4 \times 6 =$ _____

> If a rectangle is *a* units wide and *b* units long,
> its area is $a \times b$ square units.

7. What is the area of the rectangle above? _____ cm²

Find the area of each rectangle below.

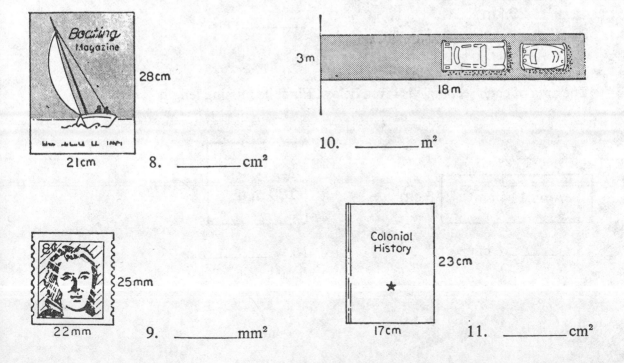

8. _____ cm²

10. _____ m²

9. _____ mm²

11. _____ cm²

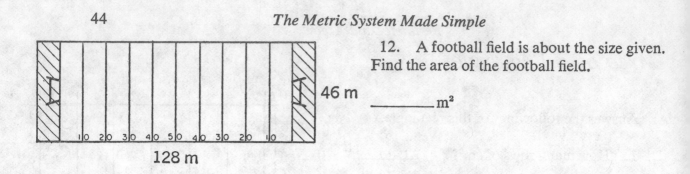

128 m

46 m

12. A football field is about the size given. Find the area of the football field.

_____ m²

13. A runway at an airport is shown. What is the area of the runway?

_____ m²

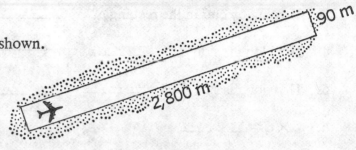

90 m

2,800 m

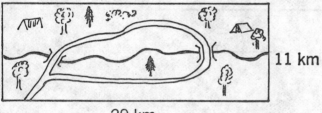

11 km

29 km

14. A rectangular state park has the size given. Find the area of the park.

_____ km²

The area of each rectangle is given below. Find the missing length.

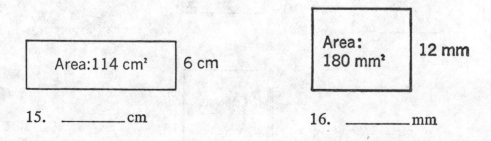

Area:114 cm²   6 cm

Area: 180 mm²   12 mm

15. _____ cm

16. _____ mm

Area of Other Geometric Shapes

Formulas for determining area

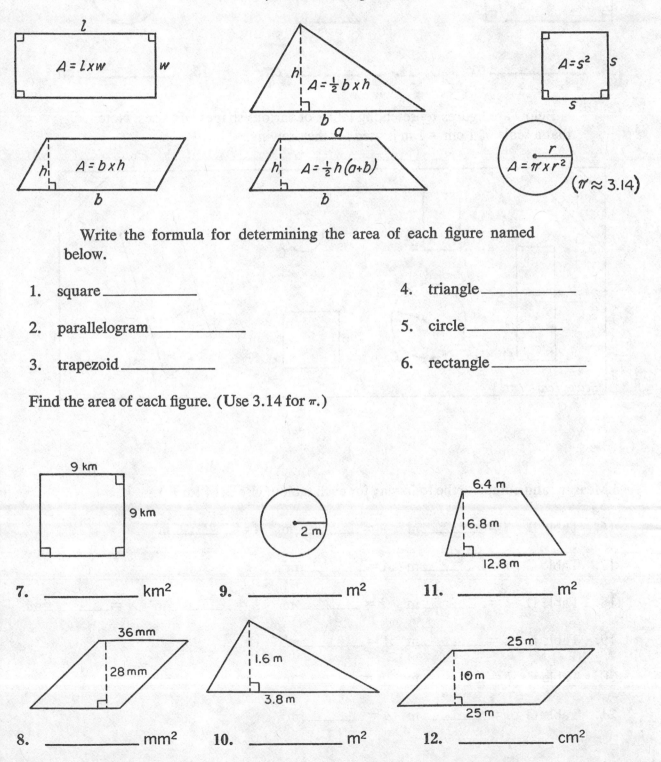

Write the formula for determining the area of each figure named below.

1. square _____

4. triangle _____

2. parallelogram _____

5. circle _____

3. trapezoid _____

6. rectangle _____

Find the area of each figure. (Use 3.14 for π.)

7. _____ km²   9. _____ m²   11. _____ m²

8. _____ mm²   10. _____ m²   12. _____ cm²

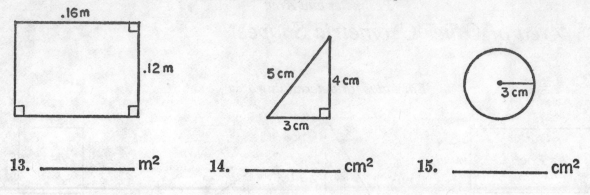

13. _____ m²   **14.** _____ cm²   **15.** _____ cm²

Below are figures representing tables of various shapes and sizes. Note that a scale of 1 cm = 1 m is used for the drawing.

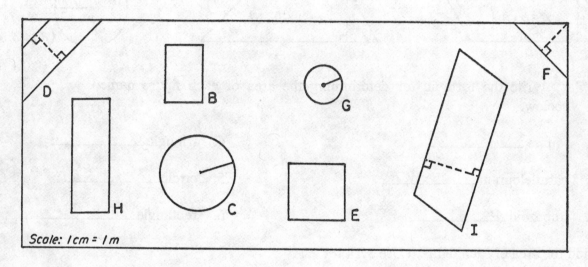

Measure and complete the following for each table. (Use 3.14 for π.)

16.   Table B $l=$ _____ m $w=$ _____ m $A=$ _____ m²

17.   Table C $r=$ _____ m $A=$ _____ m²

18.   Table D $a=$ _____ m $b=$ _____ m $h=$ _____ m $A=$ _____ m²

19.   Table E $s=$ _____ m $A=$ _____ m²

20.   Table F $b=$ _____ m $h=$ _____ m $A=$ _____ m²

21.   Table G $r=$ _____ m $A=$ _____ m²

22. Table H $l =$ _____ m $w =$ _____ m $A =$ _____ m²

23. Table I $b =$ _____ m $h =$ _____ m $A =$ _____ m²

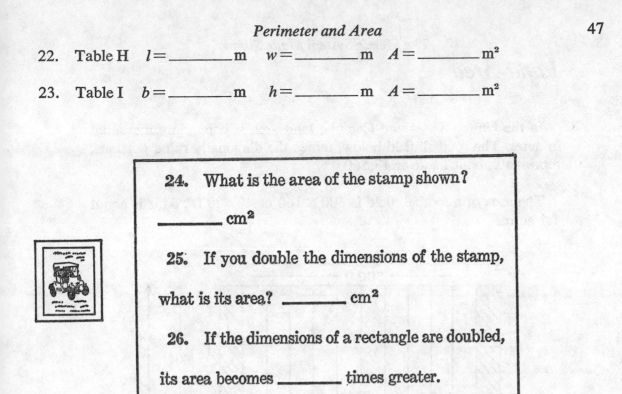

24. What is the area of the stamp shown?

_____ cm²

25. If you double the dimensions of the stamp,

what is its area? _____ cm²

26. If the dimensions of a rectangle are doubled,

its area becomes _____ times greater.

Find the area of each square shown below.

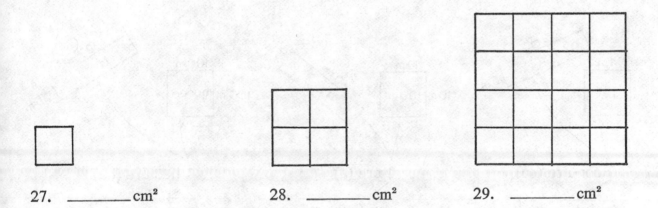

27. _____ cm²   28. _____ cm²   29. _____ cm²

30. What happens to the area of a square when you double the
length of its sides? _____

Land Area

In the United States and Canada, land area is commonly measured in acres. The football field below, minus the diagonally ruled portions, represents about 1 acre or 43,560 ft².

The area of a football field is 300×160 or 48,000 ft². This is about 1.1 acres.

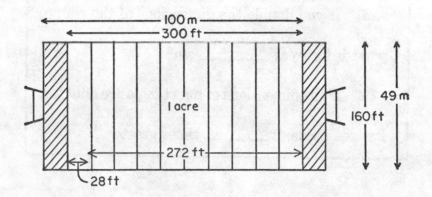

In the metric system, land area is commonly measured in **ares, centares,** and **hectares.**

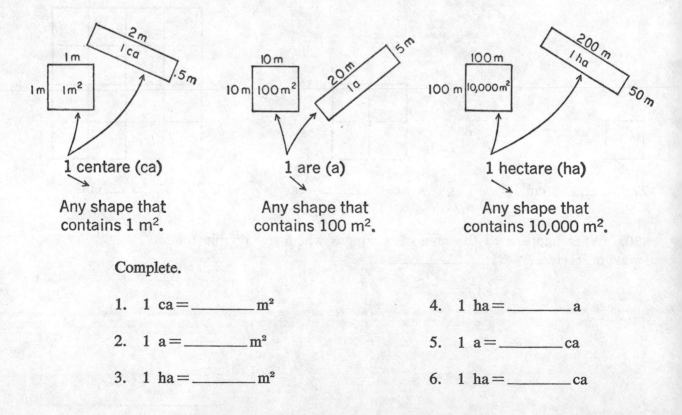

1 centare (ca)	1 are (a)	1 hectare (ha)
Any shape that contains 1 m².	Any shape that contains 100 m².	Any shape that contains 10,000 m².

Complete.

1. 1 ca = _____ m²

2. 1 a = _____ m²

3. 1 ha = _____ m²

4. 1 ha = _____ a

5. 1 a = _____ ca

6. 1 ha = _____ ca

7.   Is a centare slightly smaller or larger than a square yard? _____

8.   A football field is about 49 m by 100 m or

_____m².

9.   About how many ares are in two football fields?

_____ a

10.   Is a hectare slightly smaller or larger than two football fields?   _____

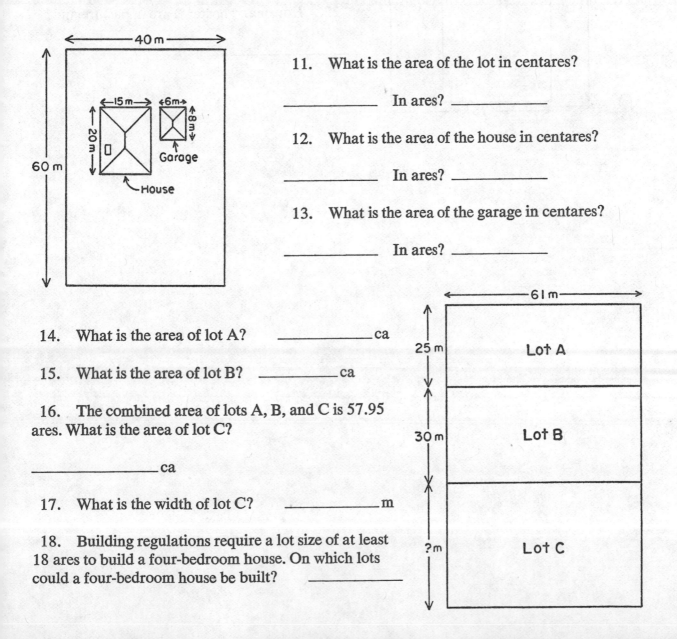

11.   What is the area of the lot in centares?

_____ In ares? _____

12.   What is the area of the house in centares?

_____ In ares? _____

13.   What is the area of the garage in centares?

_____ In ares? _____

14.   What is the area of lot A?   _____ ca

15.   What is the area of lot B?   _____ ca

16.   The combined area of lots A, B, and C is 57.95 ares. What is the area of lot C?

_____ ca

17.   What is the width of lot C?   _____ m

18.   Building regulations require a lot size of at least 18 ares to build a four-bedroom house. On which lots could a four-bedroom house be built?   _____

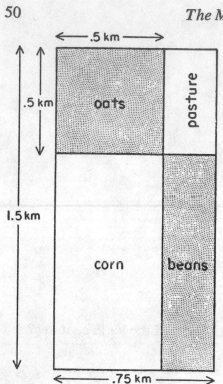

19. How many hectares are planted in oats?

_____

20. How many hectares are planted in corn?

_____

21. How many hectares are planted in beans?

_____

22. How many hectares are in pastureland?

_____

Comparing Perimeter and Area

Use a centimeter as the unit of length. Draw three different rectangles. Each rectangle is to have an area of 16 square centimeters.

Then find the perimeter of each rectangle.

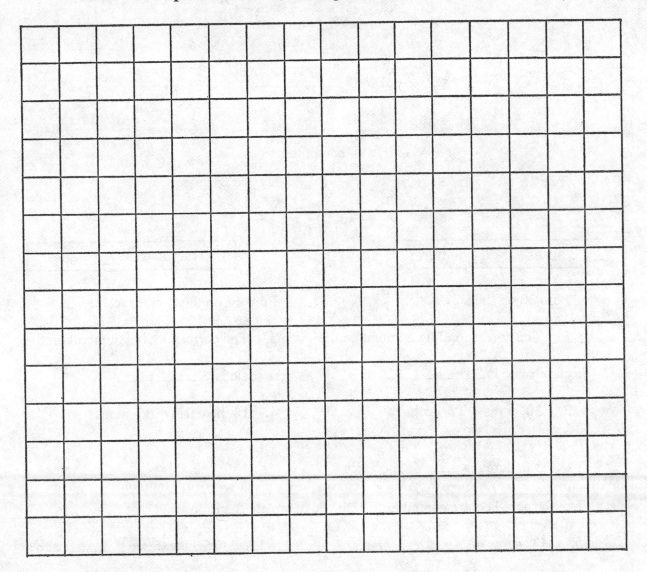

1. If two rectangles have the same area, do they have the same perimeter? _____

2. Which of your rectangles has the smallest perimeter?

_____

Conversions

1 acre ≈ .4 hectare   1 hectare ≈ 2.5 acres

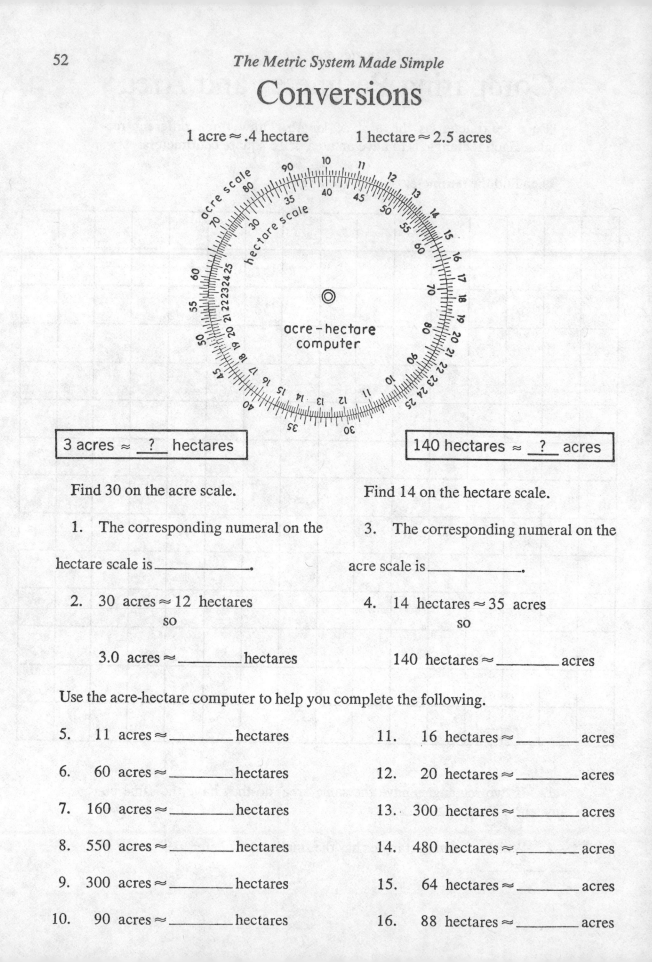

acre–hectare
computer

| 3 acres ≈ __?__ hectares | | 140 hectares ≈ __?__ acres |

Find 30 on the acre scale.   Find 14 on the hectare scale.

1. The corresponding numeral on the   3. The corresponding numeral on the

hectare scale is _____.   acre scale is _____.

2. 30 acres ≈ 12 hectares   4. 14 hectares ≈ 35 acres
            so   so

    3.0 acres ≈ _____ hectares   140 hectares ≈ _____ acres

Use the acre-hectare computer to help you complete the following.

5.   11 acres ≈ _____ hectares   11.   16 hectares ≈ _____ acres

6.   60 acres ≈ _____ hectares   12.   20 hectares ≈ _____ acres

7.   160 acres ≈ _____ hectares   13.   300 hectares ≈ _____ acres

8.   550 acres ≈ _____ hectares   14.   480 hectares ≈ _____ acres

9.   300 acres ≈ _____ hectares   15.   64 hectares ≈ _____ acres

10.   90 acres ≈ _____ hectares   16.   88 hectares ≈ _____ acres

Check Your Progress

Find the perimeter of each figure.

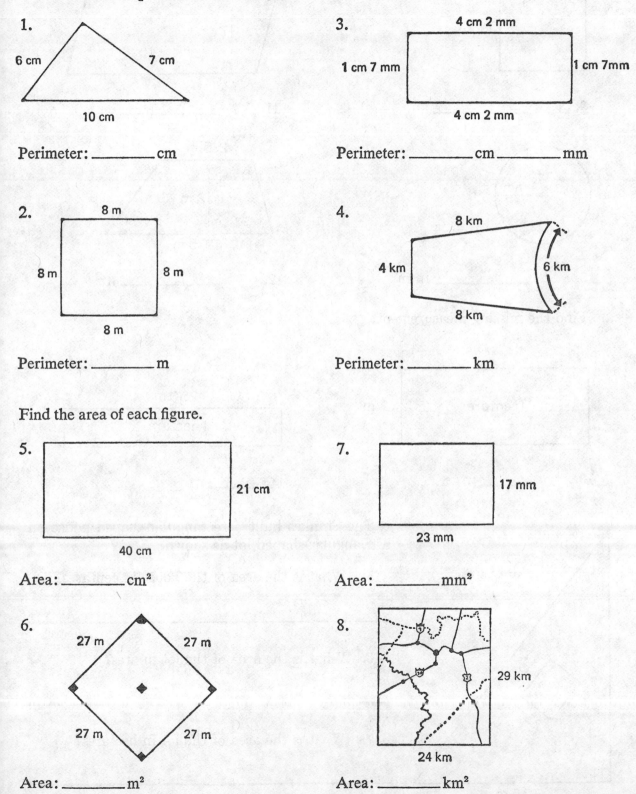

1.

6 cm   7 cm

10 cm

Perimeter: _____ cm

2.

8 m

8 m   8 m

8 m

Perimeter: _____ m

3.

4 cm 2 mm

1 cm 7 mm   1 cm 7mm

4 cm 2 mm

Perimeter: _____ cm _____ mm

4.

8 km

4 km   6 km

8 km

Perimeter: _____ km

Find the area of each figure.

5.

21 cm

40 cm

Area: _____ cm²

6.

27 m   27 m

27 m   27 m

Area: _____ m²

7.

17 mm

23 mm

Area: _____ mm²

8.

29 km

24 km

Area: _____ km²

The Metric System Made Simple

Find the area of each figure. (Use 3.14 for π.)

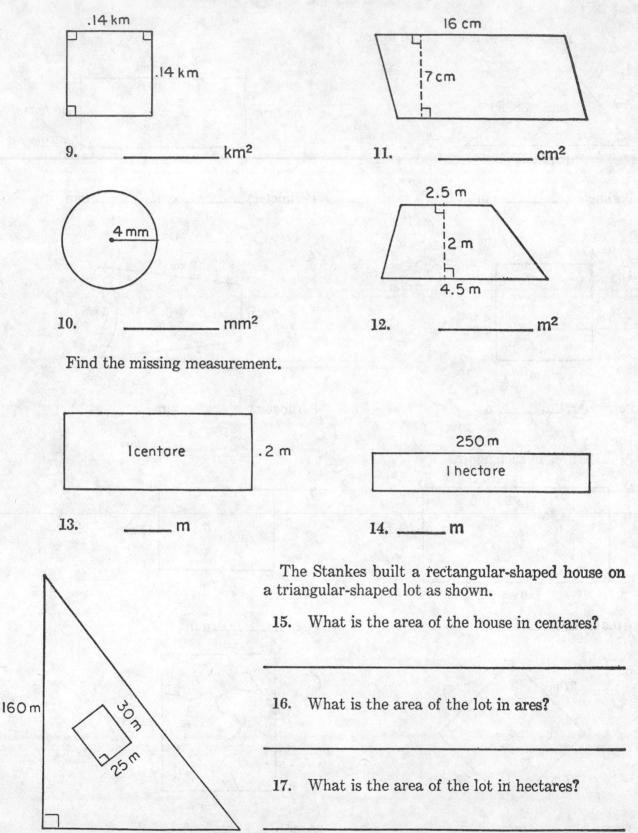

.14 km

.14 km

9. _____ km²

16 cm

7 cm

11. _____ cm²

4 mm

10. _____ mm²

2.5 m

2 m

4.5 m

12. _____ m²

Find the missing measurement.

1 centare

.2 m

13. ____ m

250 m

1 hectare

14. ____ m

The Stankes built a rectangular-shaped house on a triangular-shaped lot as shown.

15. What is the area of the house in centares?

_____

16. What is the area of the lot in ares?

_____

17. What is the area of the lot in hectares?

_____

160 m

30 m

25 m

125 m

Volume and Capacity

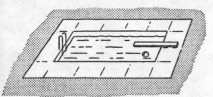

How much water?

Volume

Fuel capacity: *82 liters*
Piston displacement: *2,778 cc*

Cubic Units

Cubic units are used to measure volume.

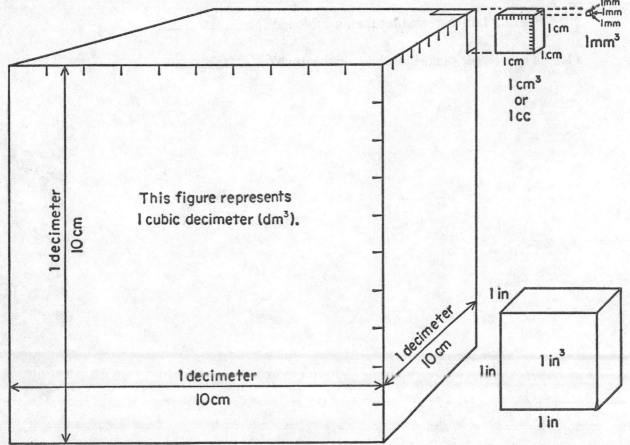

This figure represents
1 cubic decimeter (dm³).

1 decimeter
10 cm

1 decimeter
10 cm

1 decimeter
10 cm

1 mm
1 mm
1 mm
1 cm
1 mm³

1 cm
1 cm
1 cm³
or
1 cc

1 in
1 in
1 in³
1 in

You can abbreviate *cubic centimeter* as **cm³.**

1. You can write *cubic millimeter* as _____.

2. You can write *cubic decimeter* as _____.

3. You can also write *cubic centimeter* as _____.

Complete each problem below.

4. 1 cm = _____ mm

5. 1 cm^2 = _____ mm^2

6. 1 cc = _____ mm^3

7. 1 dm = _____ cm

8. 1 dm^2 = _____ cm^2

9. 1 dm^3 = _____ cc

10. Is a dm^3 larger or smaller than a cubic inch? _____

11. Is a cc larger or smaller than a cubic inch? _____

Finding Volume

To find the volume of a figure like that shown below, think of separating it into layers.

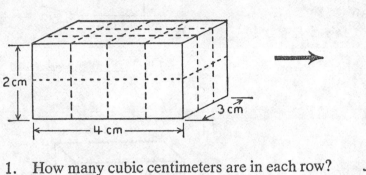

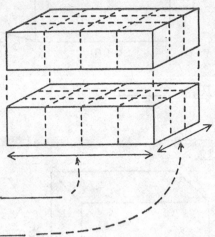

1. How many cubic centimeters are in each row?   _____

2. How many rows are there in each layer?   _____

3. $4 \times 3 =$ _____

4. How many cubic centimeters are in each layer?   _____

5. How many layers are there?   _____

6. $12 \times 2 =$ _____

7. How many cubic centimeters are in both layers?   _____

> If a figure is l units long, w units wide, and h units high,
> its volume is $l \times w \times h$ cubic units.

8. What is the volume of the figure?   _____ cc

9. $4 \times 3 \times 2 =$ _____

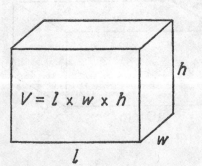

Assume the length is 10 m, the width 5 m, and the height 8 m. The volume of such a rectangular solid can be found as follows.

$$V = l \times w \times h$$
$$= 10 \times 5 \times 8$$
$$= 400$$

The volume of the rectangular solid is   400   m³.

Find the volume of each rectangular solid below.

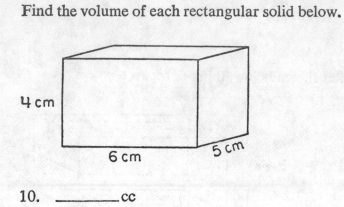

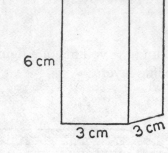

10. _____ cc

11. _____ cc

12.

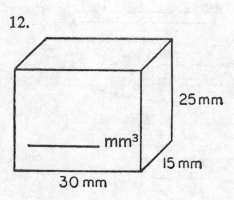

13.

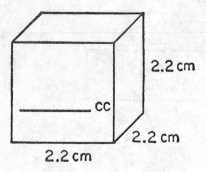

14.

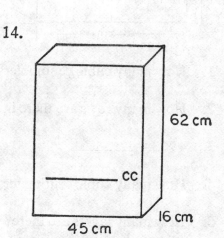

15.

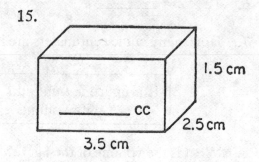

16.

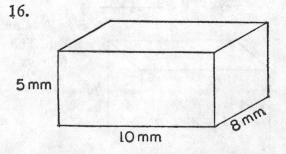

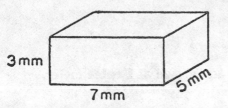

17. _____ cu mm

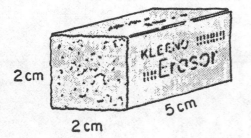

18. The picture at the left shows an eraser. Find its volume.

_____ cc

19. A pack of address labels is shown at the right. Find its volume.

_____ cu mm

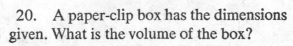

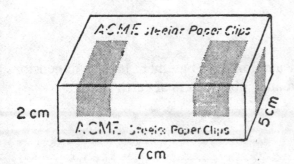

20. A paper-clip box has the dimensions given. What is the volume of the box?

_____ cc

21. An aquarium has the dimensions given. How much water can it hold?

_____ cc

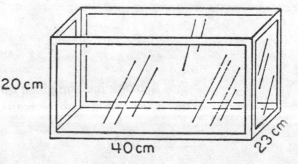

Larger Units of Volume

This page is not large enough to show the actual size of **1 cubic meter.** You can write *1 cubic meter* as **1 cu m** or as **1 m³.**

1.   Each edge of a cubic meter is _____ m long.

2.   Each edge of a cubic meter is _____ dm long.

3.   1 m³ = _____ dm³

4.   1 m³ = _____ cc

5.   Find the volume of the figure below.

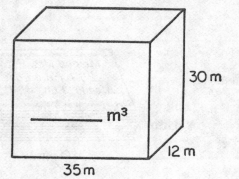

30 m

_____ m³

12 m

35 m

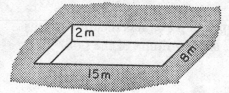

6.   An excavation for a basement has the dimensions given. How many cubic meters of earth were removed?   _____ m³

7.   A trunk has the given dimensions. What is the volume of the trunk?

_____ dm³

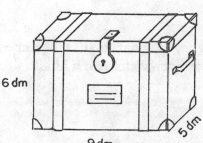

6 dm

9 dm

5 dm

8.   A car engine has 4 pistons. Each piston has a displacement of 348 cubic centimeters. What is the displacement of all 4 pistons?   _____ cc

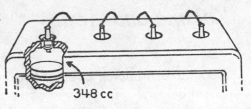

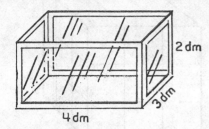

2 dm

3 dm

4 dm

9.   Given the dimensions of an aquarium as shown, find the volume of the aquarium in cubic decimeters.   _____ dm³

In cubic centimeters.   _____ cc

10.   The dimensions of a classroom are given. Find the volume of the room.   _____ m³

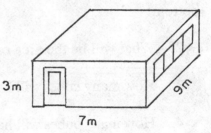

3 m

9 m

7 m

11.   In problem *10*, 20 students and 1 teacher use the classroom. How many cubic meters of space are there for each person?   _____ m³

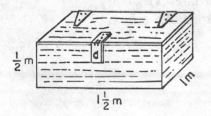

$\frac{1}{2}$ m

1 m

$1\frac{1}{2}$ m

12.   A storage box has the given dimensions [1 m = 10 dm, so ½ m = (½ × 10) dm.]. Find the volume of the box in cubic decimeters.

_____ dm³

Volume and Area

Wooden centimeter cubes were stacked in order to build the figure shown below.

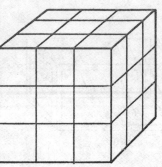

1. How many cubes were needed?   _____

If one wanted to paint the outside of the figure, including the top and bottom:

2. What will be the area of the painted surface?   _____ cm²

3. How many cubes will have only 3 faces painted?   _____

4. How many cubes will have only 2 faces painted?   _____

5. How many cubes will have only 1 face painted?   _____

6. How many cubes will have no faces painted?   _____

Get yourself 24 cubes (sugar cubes, dice, or wooden blocks). You can stack them in six different ways to build a figure shaped like a box. One of these figures is shown below. Build the others.

Area:

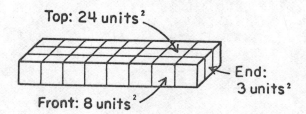

Top: 24 units²

Front: 8 units²

End: 3 units²

Top and bottom:	(2 × 24) or	48 units²
Both ends:	(2 × 3) or	6 units²
Front and back:	(2 × 8) or	16 units²
Total Area:		70 units²

Find the volume and the area of each figure.

7. If two figures have the same volume, do they have the same

area?   _____

8. What are the dimensions of the figure with the least area?

_____

9. What are the dimensions of the figure with the largest area?

_____

Capacity

The volume of a container tells you these two things.

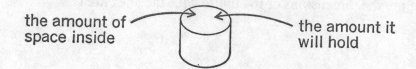

the amount of space inside   the amount it will hold

The amount a container will hold is called its **capacity.**

Capacity can be given in units of liquid measure or in cubic units.

The capacity, in liquid measure, of 1 cubic decimeter is *1 liter* (pronounced *lēt-er*).

$$1 \text{ liter} = 1 \text{ dm}^3$$

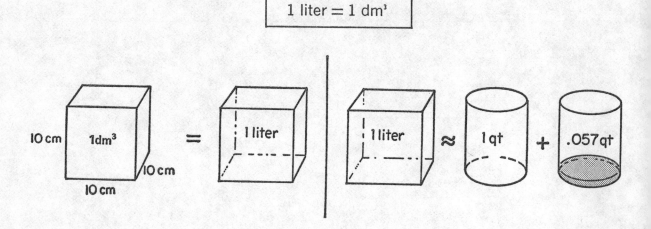

1 dm³ = 1 liter   1 liter ≈ 1.057 quarts

1. Is a liter slightly more, slightly less,

or the same amount as a dm³?   _____

2. 1 dm³ = _____ cc

3. 1,000 cc = _____ liter

4. Is a liter slightly more, slightly less,

or the same amount as a quart?   _____

5. 1.057 qt ≈ 1 liter or _____ cc

6. 1 qt ≈ 1,000 ÷ 1.057 or _____ cc

7. Could you put 1 quart of milk in a liter bottle? _____

8. Could you put 1 liter of milk in a quart bottle? _____

9. Could you put 4 liters of milk in a gallon bottle? _____

How do you know? _____

The metric prefix most commonly used with liter is *milli*, meaning .001.

1 liter = 1,000 milliliters (ml)   1 ml = .001 liter

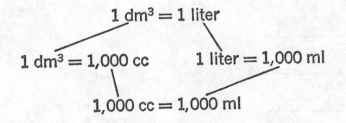

10.   1 cc = _____ ml

11. 7.54 liters = _____ ml   15. 3,450 ml = _____ liters

12. .456 liter = _____ ml   16. 78.5 ml = _____ liter

13. 65 liters = _____ ml   17. 5 ml = _____ liter

14. .5 liter = _____ ml   18. 250 ml = _____ liter

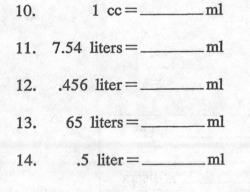

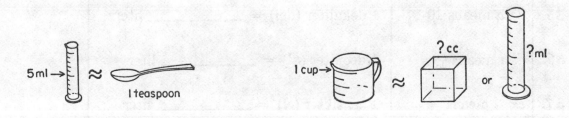

48 teaspoons = 1 cup

19.   5 ml ≈ _____ teaspoon   23.   1 cup = _____ teaspoons

20.   1 ml = _____ cc   24.   1 cup ≈ (48 × 5) or _____ cc

21.   5 ml = _____ cc   25.   1 cup ≈ _____ ml

22.   1 teaspoon ≈ _____ cc   26.   .75 cup ≈ _____ ml

The prefix *deka,* meaning 10, is often used with liter.

1 dekaliter (dal) = 10 liters

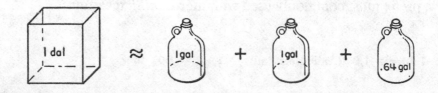

27.   1 liter = _____ cc

28.   1 dal = _____ liters

29.   1 dal = _____ cc

30.   1 dal = _____ ml

31.   1 dal ≈ _____ gal

32.   1 gal ≈ 10,000 ÷ 2.64 or _____ ml

33. *kilo* means 1,000	1 kiloliter (kl) = _____ liters
34. *hecto* means 100	1 hectoliter (hl) = _____ liters
35. *deka* means 10	1 dekaliter (dal) = _____ liters
36. *deci* means $\frac{1}{10}$	1 deciliter (dl) = _____ liter
37. *centi* means $\frac{1}{100}$	1 centiliter (cl) = _____ liter
38. *milli* means $\frac{1}{1000}$	1 milliliter (ml) = _____ liter

Drink! Drink! Drink!

39. Which quantity of your favorite drink would you

choose? _____

(P.S. Choose the greatest amount possible.)

    A   3 dal
    B   30 gal
    C   300 liters
    D   3,000 cc
    E   30,000 ml

40. If you drink 1,000 cc of that quantity every day,

how many days will it last? _____

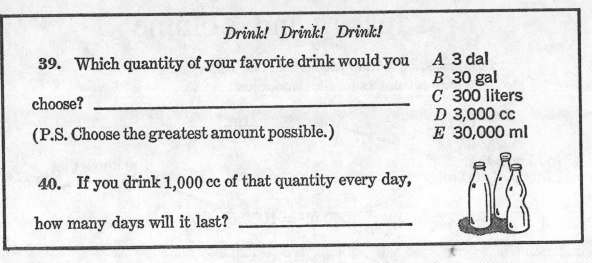

Capacity and Volume

In the preceding section you learned that 1 liter and 1 cubic decimeter are two names for the same volume.

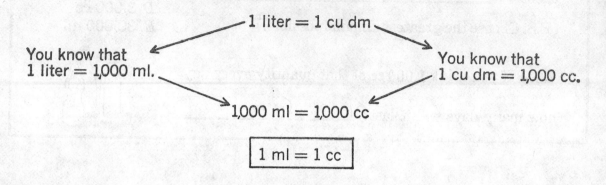

1 liter = 1 cu dm

You know that
1 liter = 1,000 ml.

You know that
1 cu dm = 1,000 cc.

1,000 ml = 1,000 cc

1 ml = 1 cc

A plastic tray has the dimensions given below.

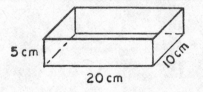

5 cm   10 cm

20 cm

1.   The volume of the tray is _____ cc.

2.   Since 1 cc = 1 ml, the capacity of the tray is _____ ml.

3.   Since 1,000 ml = 1 liter, the capacity can also be given as _____ liter.

4.   The volume of this can is _____ cc.

5.   Its capacity is _____ ml.

6.   Its capacity can also be given as _____ liters.

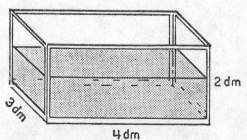

20 cm   GASOLINE

10 cm

15 cm

An aquarium has the dimensions given below. It is half filled with water.

The amount of water in the aquarium can be given in these two ways.

7.   _____ dm³

8.   _____ liters

2 dm

3 dm

4 dm

Problems About Volume and Capacity

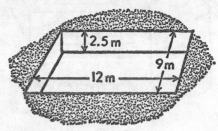

1. How much earth must be removed to make a basement excavation as indicated? _____

2. In problem *1*, it cost $1.20 per cubic meter to remove the earth. What was the cost for the excavation? _____

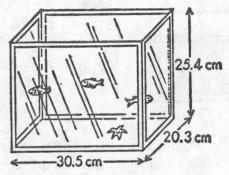

3. What is the volume in cc of the aquarium?

_____

4. How many liters of water are needed to fill the aquarium?

_____

5. A scientist is to mix the liquids shown in a single beaker. What will the volume (in ml) of the mixture be?

_____

What is the volume in liters? _____

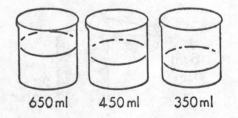

6. Andy bought 3 liters of motor oil for his car. He paid $1.77 for the oil. How much did he pay for each liter? _____

$1.77

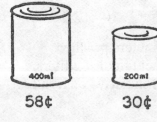

58¢   30¢

7. Tom compared two cans of fruit juice in a grocery store. A 400 ml can sold for $.58. A 200 ml can sold for $.30. Which is more economical? _____ ml can

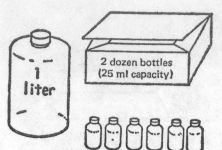

8.   A pharmacist had 1 liter of medicine which he used to fill 30 smaller bottles with 25 ml each. How much medicine did he use in the 30 bottles? _____

How much medicine was left in the liter container?

_____

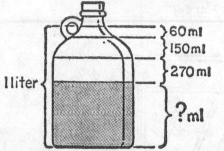

9.   Tricia's mother bought 1 liter of liquid cleaner. She used 60 ml to clean a drawer, 150 ml for a door, and 270 ml for the stove. How much liquid cleaner did she have left?

_____ ml

10.   A tank contains 48 liters of water. It has a drain pipe that will drain 500 ml per minute. How long would it take to drain the tank?

_____ hour _____ min

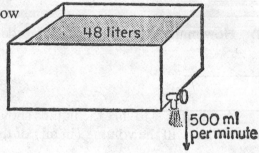

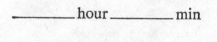

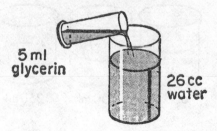

11.   In a science experiment John mixed 5 ml of glycerin with 26 cc of water. What was the volume of the resulting mixture?   _____ cc

12.   The gasoline tank on a sports car has a capacity of 60 liters. The car can travel 12 kilometers on a liter of gasoline. How far can it travel on a full tank of gasoline?   _____ km

1 liter for |—5km—|

80 liters for |—— ?km ——|

13.   Mrs. Langley's car averages 5 kilometers per liter (5km/liter) of fuel. The capacity of its fuel tank is 82 liters. Allowing 2 liters as reserve, how far could she drive on a full tank of fuel?   _____ km

14. The stainless steel ruler shown is 1 mm thick. How many mm³ of stainless steel are contained in a ruler like

this? _____

What is the amount in cc? _____

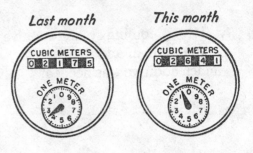

32 mm ⟸⟹   ←————— 469.2 mm —————→

15. Last month a water-meter reading was 2,175 cubic meters. This month the reading is 2,641 cubic meters. How much water was used during the month?

_____ m³

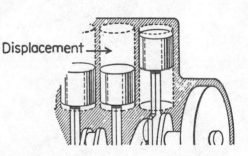

Last month   This month

16. There are 18 liters of water in an aquarium. The filter pump circulates 30 milliliters of water every minute. How long will it take to circulate all of the water? _____ hours

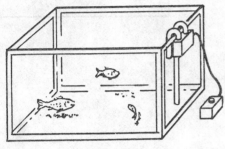

17. The total engine displacement for a car is 2,778 cc. Each of the six cylinders has the same displacement. What is the displacement of each cylinder?

_____ cc

Displacement →

18. The oil capacity of a car is shown on the chart. What is the total oil capacity for this car? _____ liters

19. Since 1 liter ≈ 1.057 quarts, 5.5 liters ≈ 5.5 × 1.057 or _____ quarts.

20. Suppose you have 5 quarts of oil. Would this be enough to change the oil in the crankcase of the car indicated on the chart? _____

21. A drum contains 20 dal of oil. Suppose it takes 6.5 liters for each oil change. How many complete oil changes can be made using the oil from this drum? _____

Oil Capacity	
crankcase	5.5 liters
oil filter	.5 liter
oil cooler	.5 liter

Fuel Capacity	
gasoline tank	2.5 dal

Radiator Capacity	
purified water	10.5 liters

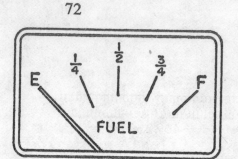

22.   The gas tank of the car indicated on the chart on the preceding page was filled to capacity. How many liters of fuel were in the tank? _____

23.   The car in problem 22 was driven 395 km before running out of fuel. How many km/liter did the car get? _____

24.   A cooling-system additive was poured into the radiator so the ratio of additive to water was 2.5 cc to 1 liter. How much additive was needed? _____

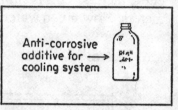

Anti-corrosive additive for ⟶ cooling system

Conversions

$$1 \text{ m}^3 = 1.3 \text{ yd}^3 \qquad 1 \text{ liter} \approx 1.057 \text{ quarts}$$

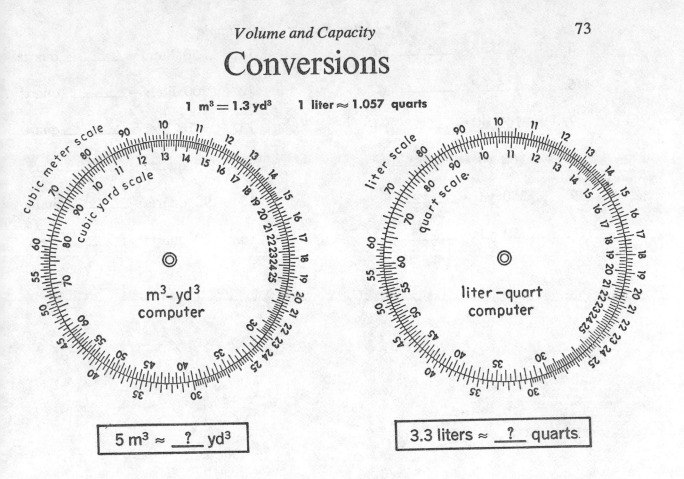

| $5 \text{ m}^3 \approx$ __?__ yd^3 | $3.3 \text{ liters} \approx$ __?__ quarts. |

Find 50 on the m³ scale.

1.   The corresponding numeral on the yd³ scale is _____.

2. $50 \text{ m}^3 \approx 65 \text{ yd}^3$
           so

       $5.0 \text{ m}^3 \approx$ _____ yd^3

Find 33 on the liter scale.

3.   The corresponding numeral on the quart scale is _____.

4. $33 \text{ liters} \approx 35 \text{ quarts}$
           so

       $3.3 \text{ liters} \approx$ _____ quarts

Use the computers above to help you complete the following.

5.   8.5 m³ ≈ _____ yd³

6.   115 m³ ≈ _____ yd³

7.   140 m³ ≈ _____ yd³

8.   160 yd³ ≈ _____ m³

9.   350 yd³ ≈ _____ m³

10.   50 yd³ ≈ _____ m³

11.   50 liters ≈ _____ quarts

12.   900 liters ≈ _____ quarts

13.   .18 liter ≈ _____ quart

14.   9.5 quarts ≈ _____ liters

15.   190 quarts ≈ _____ liters

16.   20 quarts ≈ _____ liters

Check Your Progress

Find the volume of each rectangular solid below.

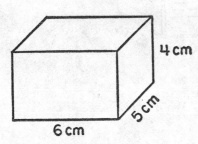

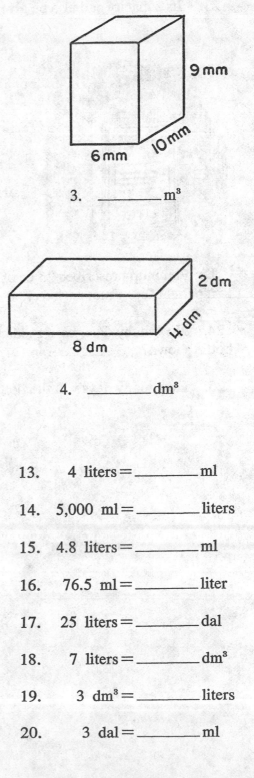

1. _____ cc

3. _____ m³

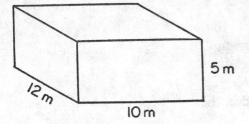

2. _____ m³

4. _____ dm³

Complete the following.

5.   1 liter = _____ dm³

6   1 liter = _____ cc

7.   1 liter = _____ ml

8.   1 cc = _____ ml

9.   1 dal = _____ liters

10.   50 cc = _____ ml

11.   9 ml = _____ cc

12.   12 cc = _____ ml

13.   4 liters = _____ ml

14.   5,000 ml = _____ liters

15.   4.8 liters = _____ ml

16.   76.5 ml = _____ liter

17.   25 liters = _____ dal

18.   7 liters = _____ dm³

19.   3 dm³ = _____ liters

20.   3 dal = _____ ml

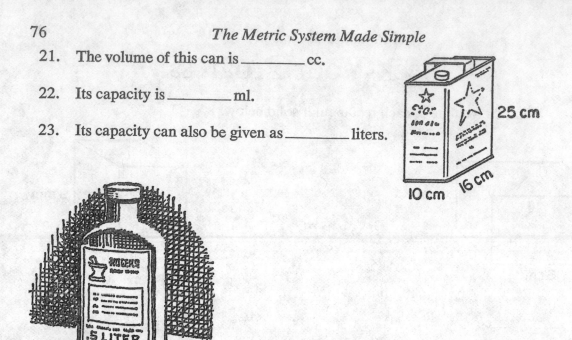

21. The volume of this can is _____ cc.

22. Its capacity is _____ ml.

23. Its capacity can also be given as _____ liters.

Dr. Fixumup prescribed 25 cc of XY drug to be given to a patient four times each day.

24. How many dosages of medicine are contained in the bottle shown?   _____

25. How many days will the bottle of medicine last?   _____

Mass and Weight

How much does it weigh?

Definitions

What is its mass?

You have probably seen astronauts "float" in space. They were weightless. **Weight** is the force of the earth's gravitational pull on an object. This force decreases as the distance between the object and earth increases. Thus, objects in space have little, if any, weight.

In the metric system, weight is measured in **grams.** You can write **1 gram as 1 g.**

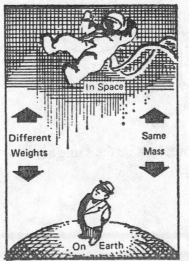

Mass is the amount of material in an object. An object is composed of the same amount of material whether it is in space or on earth. Thus, the mass of an object is the same regardless of where it may be.

On earth the mass and weight of an object are nearly alike. In fact, they are so closely related that we use the same units for both. For example, you may have a mass of 60 kilograms. If so, your weight on earth is 60 kilograms.

The **kilogram** is the standard unit of mass in the metric system. As noted below, there is but one standard for the unit of mass. However, every nation has one or more duplicates of this standard.

Kilogram

The standard for the unit of mass is the *kilogram*. The kilogram is a cylinder of platinum-iridium alloy kept by the International Bureau of Weights and Measures near Paris. A duplicate kilogram is in the custody of the National Bureau of Standards in Washington. This duplicate serves as the mass standard for the United States.

(The kilogram is the only base unit defined by a man-made product.)

Since we'll be "down to earth" in the next few lessons, no further distinction will be made between weight and mass. The term weight will be used exclusively.

1 kilogram = 1,000 grams (g)   **1 g = .001 kg**

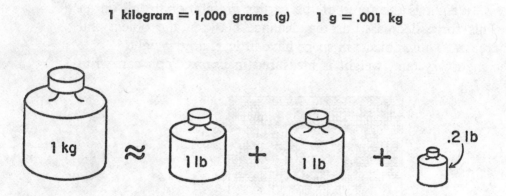

1.   Is a kilogram slightly more, less, or the same as two pounds? _____

2.   1 kg ≈ _____ lb

3.   1,000 g ≈ _____ lb

4.   1 g ≈ 2.2 ÷ 1,000 or _____ lb

5.   3 kg = _____ g

6.   .05 kg = _____ g

7. 2,500 g = _____ kg

8.   35.5 g = _____ kg

The prefix *milli,* meaning .001, is often used with grams.

$$1 \text{ g} = 1,000 \text{ milligrams (mg)} \qquad 1 \text{ mg} = .001 \text{ g}$$

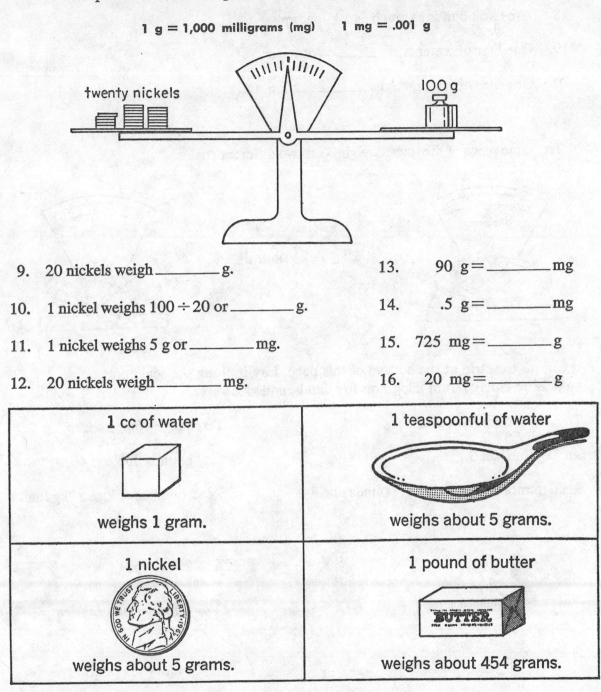

9.   20 nickels weigh _____ g.

10.   1 nickel weighs $100 \div 20$ or _____ g.

11.   1 nickel weighs 5 g or _____ mg.

12.   20 nickels weigh _____ mg.

13.   90 g = _____ mg

14.   .5 g = _____ mg

15.   725 mg = _____ g

16.   20 mg = _____ g

1 cc of water weighs 1 gram.	1 teaspoonful of water weighs about 5 grams.
1 nickel weighs about 5 grams.	1 pound of butter weighs about 454 grams.

Look at the pictures above to answer the following.

17.   How many nickels would it take to weigh 1 kilogram?

_____

18. Are two pounds as much as 1 kilogram? _____

19. One liter of water is _____ cc of water.

20. One liter of water weighs _____ kilogram.

The same piece of meat was weighed on two different scales.

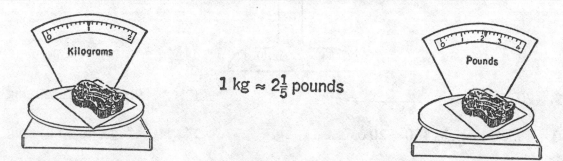

$$1 \text{ kg} \approx 2\tfrac{1}{5} \text{ pounds}$$

Cut off the strip at the bottom of this page. Lay it along the scale on page 28 to change from kilograms to pounds, and vice versa.

To change 5 kg to pounds:

Set "kg line" on 5.

Read number of pounds on "pounds line."

To change 40 pounds to kg:

Set "pounds line" on 40.

Read number of kg on "kg line."

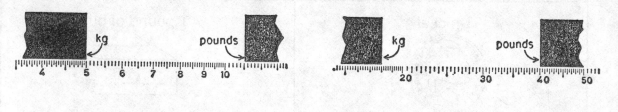

5 kg ≈ 11 pounds   18 kg ≈ 40 pounds

Complete the following.

21.   6 kg ≈ _____ pounds   24. _____ kg ≈ 20 pounds

22.   10 kg ≈ _____ pounds   25. _____ kg ≈ 80 pounds

23.   25 kg ≈ _____ pounds   26. _____ kg ≈ 35 pounds

27.   Sara weighs 72 pounds. Tom weighs 72 kilograms. Who is heavier?   _____

How do you know?   _____

28.   Max used 25 ml of water for a science experiment.

How many cubic centimeters of water did he use?   _____ cc

What was the weight of the water?   _____ g

Grams, Kilograms, and Metric Tons

$$1,000 \text{ kg} = 1 \text{ metric ton (t)}$$

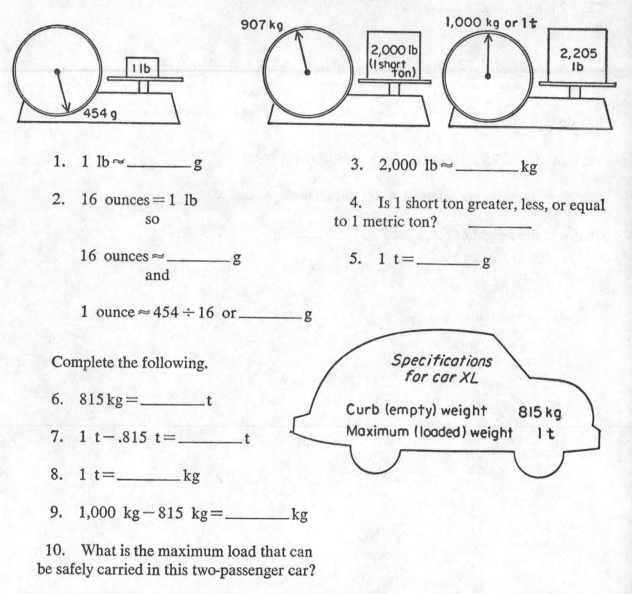

1.   1 lb ≈ _____ g

2.   16 ounces = 1 lb
         so

   16 ounces ≈ _____ g
         and

   1 ounce ≈ 454 ÷ 16 or _____ g

Complete the following.

6.   815 kg = _____ t

7.   1 t − .815 t = _____ t

8.   1 t = _____ kg

9.   1,000 kg − 815 kg = _____ kg

10.   What is the maximum load that can
be safely carried in this two-passenger car?

_____ kg or _____ t

11.   If the driver weighs 85 kg, how much
more weight can be safely placed in the car?

_____

3.   2,000 lb ≈ _____ kg

4.   Is 1 short ton greater, less, or equal
to 1 metric ton?   _____

5.   1 t = _____ g

Specifications
for car XL

Curb (empty) weight   815 kg
Maximum (loaded) weight   1 t

1 kg = .001 t

Complete the following.

12. 1,000 mg = _____ g   18. 3 g = _____ mg

13. 1,000 g = _____ kg   19. 2,000 g = _____ kg

14. 1,000 kg = _____ t   20. 5 t = _____ kg

15. 6.5 t = _____ kg   21. 2,500 kg = _____ t

16. 6.5 t = _____ g   22. 3,500 g = _____ t

17. .75 t = _____ kg   23. 14.5 kg = _____ t

In the metric system, the volume and weight of water at 4°C have a special relationship.

Think of building a tank like that shown below.

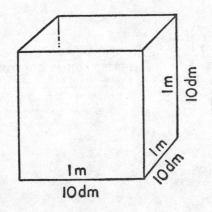

24. The volume of the tank is _____ m³.

25. The volume can also be given as _____ dm³.

Suppose you filled the tank with water.

26. Since 1 dm³ = 1 liter, it can hold _____ liters of water.

27. 1,000 liters = _____ kl

28. Since 1 liter of water weighs 1 kilogram, the water in the tank would weigh _____ kilograms.

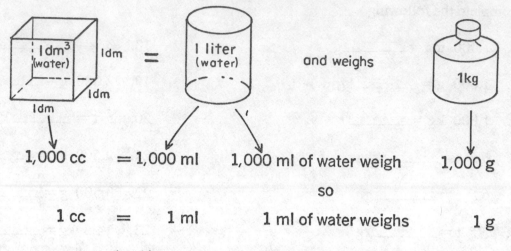

1,000 cc   = 1,000 ml   1,000 ml of water weigh   1,000 g

so

1 cc   =   1 ml   1 ml of water weighs   1 g

1 cubic meter of water weighs 1 metric ton.

Suppose this tank of water was placed on a regular scale.

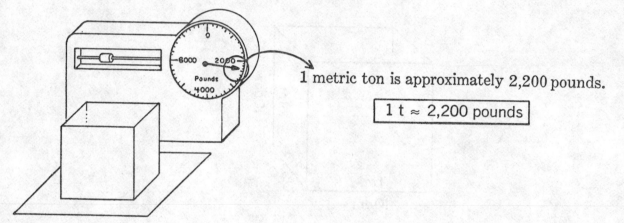

1 metric ton is approximately 2,200 pounds.

$$1 \ t \approx 2{,}200 \ \text{pounds}$$

29. 15 cc of water weigh _____ g.

30. 280 ml of water weigh _____ g.

31. 4.5 liters of water weigh _____ g.

32. 4.5 liters of water weigh _____ kg.

33. 72.5 g of water = _____ cc

34. .5 g of water = _____ ml

35. 750 g of water = _____ ml

36. 750 g of water = _____ liter

Find the weight of this water.

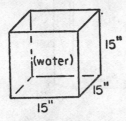

37. volume = 15 × 15 × 15 or _____ in³
     (1,728 in³ = 1 ft³)

38. volume = 3,375 ÷ 1,728 or _____ ft³
     (1 ft³ = 62.4 lb)

39. weight = 1.95 × 62.4 or _____ lb

Find the weight of this water.

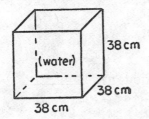

40. volume = 38 × 38 × 38 or _____ cc
     (1 cc of water weighs 1 g)

41. weight = _____ g
     (1,000 g = 1 kg)

42. weight = _____ kg

As you probably guessed, approximately the same amount of water is in each container.

43. Would you prefer to compute with the metric units or the customary units? _____ Why? _____

44.   A bag of salt pellets is shown at the right.
How many of these bags would it take to make
approximately 1 metric ton?   _____

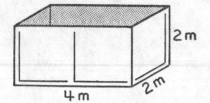

45.   A storage tank has the dimensions shown.
It is completely filled with water. What is the
weight of the water?   _____ t

46.   The weight of each player on a football team is given below.

Max	72 kg	Bill	93 kg	José	84 kg
Don	78 kg	Rex	70 kg	John	82 kg
Jeff	86 kg	Gene	75 kg	Ken	80 kg
Tony	87 kg	Randy	96 kg		

   Is their combined weight more or less than 1 metric ton? _____
By how many kilograms?   _____

47.   A turbojet traveled about 1 kilometer on 15 kilograms of fuel.
About how many metric tons of fuel were needed to travel 3,000 kilometers?   _____ t

48.   There were 30 students in one class. Their average weight was
73 pounds. Was their combined weight more or less than 1 metric
ton?   _____

Problems

Remember that 1 kg = 1,000 g and that 1 g = 1,000 mg.
Underline what you think is the approximate weight of the objects.

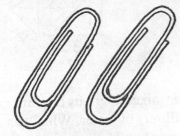

1.   1 mg   1 g   1 kg

3.   1 mg   1 g   1 kg

2.   1 mg   1 g   1 kg

4.   10 mg   10 g   10 kg

Remember that 1 kilogram is a little more than 2 pounds.
Underline what you think is the approximate weight of each object.

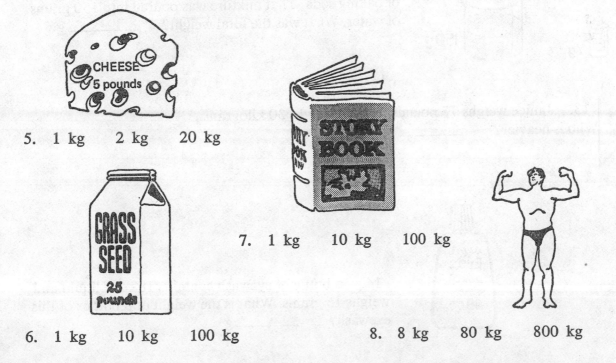

5.   1 kg   2 kg   20 kg

7.   1 kg   10 kg   100 kg

6.   1 kg   10 kg   100 kg

8.   8 kg   80 kg   800 kg

9.   A plastic box has the dimensions shown. What is the volume of the box? _____ cc   How many milliliters of water can it hold? _____ ml   What is the weight of that much water? _____ g

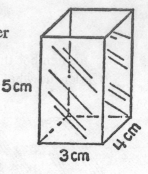

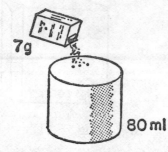

10.   A gardener mixed 7 grams of plant food with 80 milliliters of water. What was the weight of the mixture? _____ g

11.   A package of cheese weighed 3 kilograms. Arne said that was more than 6 pounds. How did he know? _____

_____

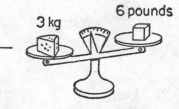

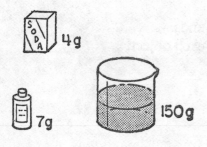

12.   Kevin mixed 7 grams of acid crystals with 4 grams of baking soda. That mixture was poured into 150 grams of water. What was the total weight? _____ g

13.   Janice weighs 72 pounds. Harold weighs 30 kilograms. Who is heavier? _____

14.   A bottle of eyewash contains 18 cc of liquid and weighs 15 grams. What is the weight of 3 bottles of this eyewash? _____ g

15. What was the total weight of the team before the game? _____ kg   After? _____ kg

Team Weight		
player	*before game*	*after game*
A	72.6 kg	70.3 kg
B	63.5 kg	60.3 kg
C	77.1 kg	73.4 kg
D	68.0 kg	65.3 kg
E	59.0 kg	57.6 kg

16. What was the difference between the team's weight before and after the game? _____

17. Which team member lost the most weight during the game? _____ How much? _____

18. Which team member lost the least weight during the game? _____ How much? _____

19. If the mouse ate the cheese, how much would the mouse weigh? _____

20. If the cat ate the mouse that ate the cheese, how much would the cat weigh? _____

21. If the man held the cat that ate the mouse that ate the cheese, how much would the combined weight be? _____

22. Which fruit has more protein and by how much?

_____

23. How many apples would contain the same amount of protein as 1 banana? _____

24. How many apples would contain the same amount of calcium as 1 banana? _____

fruit	*protein*	*calcium*
apple	.6 g	12 mg
banana	2.4 g	16 mg

25. There are 20 slices of the same size and weight in this loaf of bread. How much does each slice of bread weigh? _____

What is the weight of 3 slices? _____

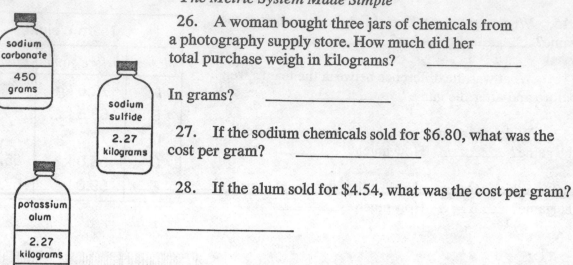

26.   A woman bought three jars of chemicals from a photography supply store. How much did her total purchase weigh in kilograms? _____

In grams?   _____

27.   If the sodium chemicals sold for $6.80, what was the cost per gram?   _____

28.   If the alum sold for $4.54, what was the cost per gram?

_____

29.   Each tablet of a drug contains 7.5 mg of an antibiotic. How many tablets can be made from the amount of antibiotic shown?   _____

30.   If each tablet produced in the preceding problem weighs 75 mg, what is the total weight of the tablets in grams?

_____

In kilograms?   _____

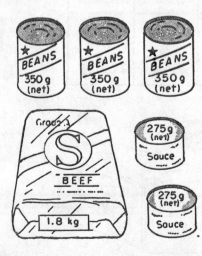

31.   In addition to the ingredients shown, a cook claims to have used exactly 500 mg of spices to create his famous *chilimess*. What is the total weight of the ingredients in grams? _____   In kilograms? _____

32.   If there were four equal servings of 225 g each (and no requests for seconds), how much of the chilimess would be left?   _____

33. An empty container is weighed, then filled with water and weighed again. How much did the water weigh in g? _____

In kg? _____ What is the capacity of the container in ml? _____ In liters? _____

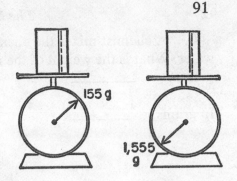

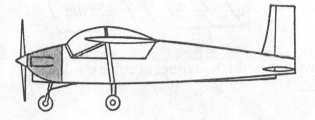

specifications for model 150	
gross weight	726 kg
empty weight	484 kg
fuel capacity	95 liters
oil capacity	5.5 liters

34. Fuel for this two-seated plane weighs .7 kg/liter. What is the weight of the fuel if the tanks are full? _____kg

35. Oil weighs .9 kg/liter. What is the weight of the oil if the crankcase is filled to capacity? _____kg

36. What is the total weight of the plane, fuel, and oil?
_____kg

37. If the pilot weighs 77 kg, what can the weight of the other passenger be, provided the fuel and oil are filled to capacity? (*Note:* The empty weight of the plane plus the weight of the fuel, oil, pilot, and passenger cannot exceed the specified gross weight.)

_____

38. In light planes like the one shown here, approximately 7 kg of air are required to completely burn .5 kg of fuel. How much air would be required to burn all the fuel in the light plane described here, provided the tanks are full? _____

39. How much fuel could be completely burned with 1 metric ton of air? _____

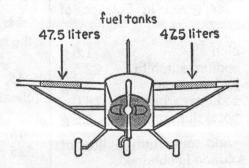

fuel tanks
47.5 liters   47.5 liters

40.   A chemist mixes the chemicals shown with 25 g of water. What is the weight of the resulting mixture in milli-grams?   _____

In grams?   _____

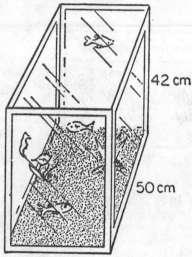

41.   What is the volume of this aquarium in cubic centi-meters?   _____

What is the capacity of the aquarium in liters?   _____

42.   Without filtration, you need a minimum of 3.5 liters to support every 25 mm length of fish. How many mm of fish could you keep in this unfiltered aquarium (provided you keep it filled with water and feed the fish regularly)?   _____

43.   What weight of water can this beaker contain in milli-liters?   _____   In liters?   _____

In grams?   _____   In kilograms?   _____

developer solution (1 liter)

Mix the chemicals in the order given in 500 cc of water.

elon	1.5 g
sodium sulphite	22.5 g
hydroquinone	6.3 g
sodium carbonate	15.0 g
potassium bromide	1.5 g

Add water until 1 liter of solution is obtained.

44.   How many grams of chemicals are mixed together in the solution?   _____

45.   What is the weight of the 500 cc of water and the chemicals?   _____

46.   The 500 cc of water and chemicals make a solution of 525 cc. How many ml of water must be added to complete the solution?   _____

Density and Specific Gravity

The **density** of any substance is the weight measure of any sample divided by the volume measure of that sample.

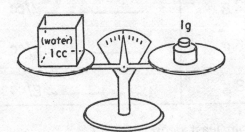 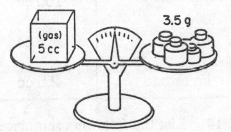

1. 1 cc of water weighs _____ g.

2. $1 \div 1 =$ _____

3. The density of water is _____ gram per cubic centimeter (g/cc).

4. 5 cc of gasoline weigh _____ g.

5. $3.5 \div 5 =$ _____

6. The density of gasoline is _____ g/cc.

Complete.

7. Which is heavier, 1 cc of water or 1 cc of gasoline?

_____

How much heavier? _____

8. If you put gasoline and water in the same container, which substance would settle to the bottom? _____ Why? _____

Complete the table by finding the density of each substance in g/cc.

	substance	volume	weight	density
9.	aluminum	8 cc	21.6 g	_____ g/cc
10.	gold	3 cc	57.9 g	_____ g/cc
11.	ice	100 cc	92 g	_____ g/cc
12.	iron	5 cc	39.5 g	_____ g/cc
13.	limestone	2.5 cc	8 g	_____ g/cc

14. List the following substances in order from least dense to most dense: *aluminum, gasoline, gold, ice, iron, limestone,* and *water.*

_____

density of gasoline ⟶ .7 g/cc

density of water ⟶ 1 g/cc

$\frac{.7}{1}$ or .7   specific gravity of gasoline

> The **specific gravity** of a substance is the ratio of the density of a substance to the density of water at 4°C.

15. 1 cc of water weighs 1 g
            so

     1,000 cc of water weigh _____ g,
            and

     1 liter of water weighs _____ g.

16. 1 cc of gasoline weighs .7 g
            so

     1,000 cc of gasoline weigh _____ g,
            and

     1 liter of gasoline weighs _____ g.

The specific gravity of a substance is numerically equal to the density of that substance when expressed in g/cc. For example, the density of carbon is 3.5 g/cc. Thus, the specific gravity of carbon is 3.5.

Complete the following table.

	substance	density	specific gravity	weight of 1 liter (1,000 cc)
17.	water	1 g/cc	1	1,000 g
18.	barium	3.5 g/cc	_____	_____
19.	calcium	1.6 g/cc	_____	_____
20.	magnesium	1.7 g/cc	_____	_____
21.	nickel	8.9 g/cc	_____	_____

A Simple Experiment

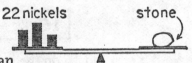

Nickels (5 g ≈ 1 nickel) and an improvised balance scale can be used to measure the weight of a stone in grams.

Fill a jar with water and set it in a pan. Slowly lower the stone into the jar. Then use a cylinder graduated in cc or ml (since 1 cc = 1 ml) to measure the volume of water that overflows into the pan. This equals the volume (in cc) of the stone.

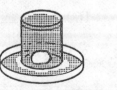

22. The stone weighs 22 nickels or _____ g.

23. The volume of the stone is _____ cc.

24. The density of the stone is _____ g/cc.

Conversions

1 kg ≈ 2.2 lb   1 metric ton (t) ≈ 1.1 short ton (T)

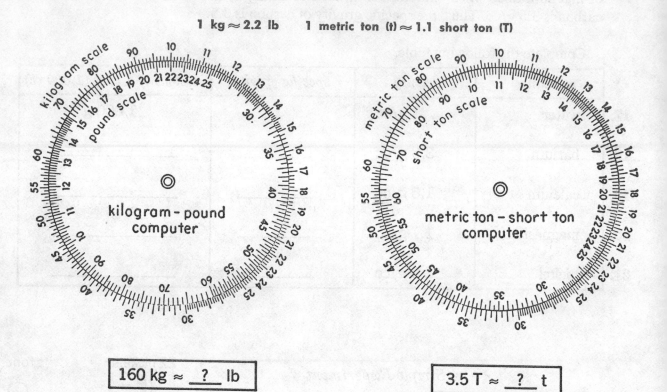

| 160 kg ≈ __?__ lb | 3.5 T ≈ __?__ t |

Find 16 on the kg scale.   Find 35 on the short ton (T) scale.

1. The corresponding numeral on the   3. The corresponding numeral on the
lb scale is _____   metric ton (t) scale is _____

2.   16 kg ≈ 35 lb   4.   35 T ≈ 32 t
      so   so

    160 kg ≈ _____ lb   3.5 T ≈ _____ t

5.   3.2 kg ≈ _____ lb   11.   400 T ≈ _____ t

6.   41 kg ≈ _____ lb   12.   .45 T ≈ _____ t

7.   180 kg ≈ _____ lb   13. 5,100 T ≈ _____ t

8.   .46 lb ≈ _____ kg   14.   520 t ≈ _____ T

9.   25 lb ≈ _____ kg   15. 1,800 t ≈ _____ T

10.   395 lb ≈ _____ kg   16.   140 t ≈ _____ T

Check Your Progress

Complete the following.

1. 4,500 mg = _____ g

2.   .05 g = _____ mg

3.   1 g = _____ mg

4.   4 g = _____ mg

5.   1,500 g = _____ kg

6.   1 kg = _____ g

7.   7 kg = _____ g

8.   75 kg = _____ g

9.   500 kg = _____ t

10.   3 t = _____ kg

11.   .65 t = _____ kg

12.   1 t = _____ kg

A can has the dimensions shown.

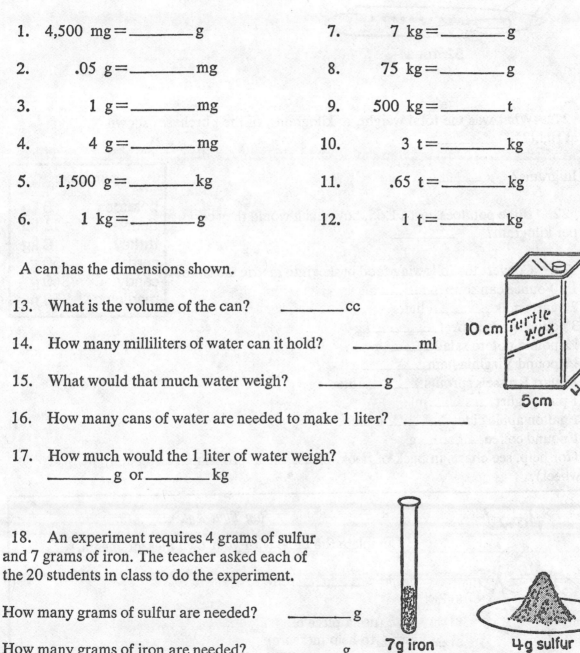

13. What is the volume of the can? _____ cc

14. How many milliliters of water can it hold? _____ ml

15. What would that much water weigh? _____ g

16. How many cans of water are needed to make 1 liter? _____

17. How much would the 1 liter of water weigh?
     _____ g or _____ kg

10 cm Turtle Wax
5 cm
4 cm

18. An experiment requires 4 grams of sulfur and 7 grams of iron. The teacher asked each of the 20 students in class to do the experiment.

How many grams of sulfur are needed? _____ g

How many grams of iron are needed? _____ g

7g iron   4g sulfur

19. A doctor prescribed some pills. Each pill contains 5 milligrams of a drug. How many of these pills can be made from 3 grams of the drug? _____

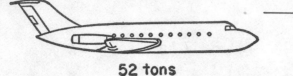

1½ tons

52 tons

20.   A small, single-engine airplane weighs 1½ metric tons. A jet airliner weighs 52 metric tons. How much heavier is the jet airliner?

_____ t

21.   What was the total weight, in kilograms, of the purchases shown at right?   _____

In grams?   _____

22.   If the potatoes cost $1.65, how much would they cost per kilogram?   _____

purchases	
potatoes	5 kg
flour	11 kg
turkey	8 kg
cereal	250 g
candy	340 g
medicine	550 mg

23.   Convert the following food basket into metric measure:
1 10-ounce can soup _____ ml
2 quarts milk _____ liters
3 pounds ground beef _____ kg
½ pound potato salad _____ g
¼ pound Virginia ham _____ g
1 quart Brussels sprouts _____ liter
1 pint yogurt _____ ml
1 gallon apple cider _____ liters
1 pound coffee _____ g
(for help, see charts in back of book, or use the appropriate conversion wheel).

Wow!

152 cm

81 cm
81 cm
81 cm

85 kg

Say That Again!

Complete the table with your own measurements.

height _____ cm

girth (use a piece of string to help measure)

chest _____ cm   *waist* _____ cm   *hips* _____ cm

mass _____ kg

Other Units

Prefixes for Metric Units

In previous sections the prefixes *kilo, hecto, deka, deci, centi,* and *milli* were applied to different units.

Other prefixes have been agreed upon for still larger and smaller units. The prefixes given below can be applied to all metric units.

prefix	symbol	meaning
tera	T	1,000,000,000,000
giga	G	1,000,000,000
mega	M	1,000,000
kilo	k	1,000
hecto	h	100
deka	da	10
deci	d	.1
centi	c	.01
milli	m	.001
micro	μ	.000 001
nano	n	.000 000 001
pico	p	.000 000 000 001
femto	f	.000 000 000 000 001
atto	a	.000 000 000 000 000 001

1. 1 megaton = _____ tons

2. 1 microsecond = _____ second

3. 1 kilocycle = _____ cycles

4. 1 megacycle = _____ cycles

A radar transmits short pulses of radio waves toward an object. Between pulses it "listens" for the radio waves that have been reflected by the object.

Radio waves travel about 186,000 miles per second.

5. Radio waves travel about _____ miles per microsecond.

6. Radio waves travel about _____ kilometers per microsecond.

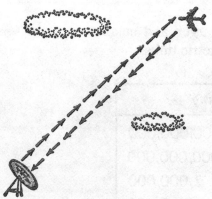

7. A radar antenna is beamed at a rain cloud. It takes 86 microseconds for the waves to travel to the cloud and back. How far is the cloud from the radar antenna? (*Hint:* rate × time = distance) _____ km

8. An airplane is 42 km from a radar station. How long will it take the radio waves to travel to the airplane and back? _____ microseconds

9. There are about 1,234 teraliters of water on the earth. Only 3% of this is fresh water. About how many teraliters of fresh water are there on the earth?

_____

10. In most cities people pay about $.06 a ton for water. At that rate, what is the cost of 1 megaton of water? _____

11. Every minute about 40 billion tons of hydrogen atoms are fused on the sun's surface.

40 billion tons = _____ megatons

40 billion tons = _____ gigatons

12. The energy in a pound of heavy hydrogen is equal to the heat produced by 5,000 tons of coal. How many megatons of coal are needed to equal the energy in 2,000 pounds of heavy hydrogen?

_____

Temperature

Celsius Scale

A doctor might say your body temperature is 98.6°. A Chicago weatherman might predict tonight's low temperature at 15°. Both of these persons would be using the **Fahrenheit (F)** scale for measuring temperature.

The **Celsius (C)** or centigrade scale for temperature is commonly used with the metric system.

Compare the two temperature scales shown below.

Between the freezing and boiling temperatures of water, the ratio of F units to C units is

$$\frac{180}{100} \text{ or } 1.8.$$

This means that a change of 1.8° F corresponds to a change of 1° C.

A change of 5° C corresponds to a change of (1.8×5) or 9° F. A change of n° C corresponds to a change of $(1.8n)$° F.

Since 32° F = 0° C, the Fahrenheit scale has a 32° "head start" on the Celsius scale. To change a Celsius reading (c) to a Fahrenheit reading (f), we must add 32 to the resulting Celsius reading.

$$f = 1.8c + 32$$

Solving this equation for c, we have

$$f - 32 = 1.8c \text{ or } \frac{f - 32}{1.8} = c.$$

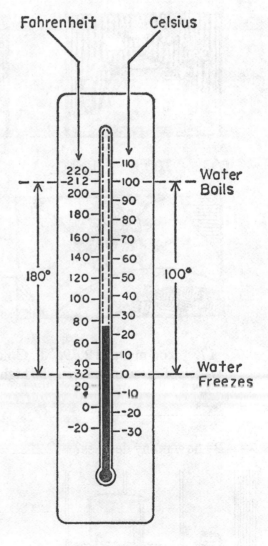

75°C = ___?___ °F

$f = (1.8 \times 75) + 32$

$f = 167$

75°C = 167°F

68°F = ___?___ °C

$c = \dfrac{68 - 32}{1.8}$

$c = 20$

68°F = 20°C

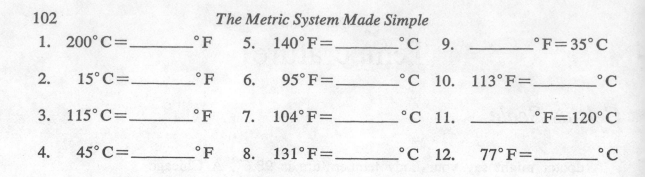

1. 200° C = _____ ° F   5. 140° F = _____ ° C   9. _____ ° F = 35° C

2.   15° C = _____ ° F   6.   95° F = _____ ° C   10. 113° F = _____ ° C

3. 115° C = _____ ° F   7. 104° F = _____ ° C   11. _____ ° F = 120° C

4.   45° C = _____ ° F   8. 131° F = _____ ° C   12.   77° F = _____ ° C

In each of the following, underline one measurement as your guess for the temperature of the water.

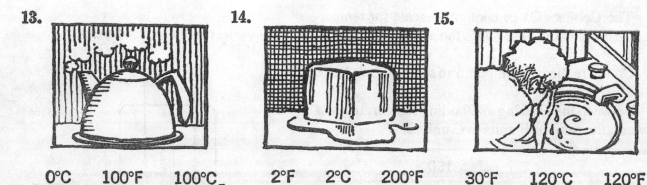

13.   **14.**   **15.**

0°C   100°F   100°C   2°F   2°C   200°F   30°F   120°C   120°F

16.   During an illness Rod's temperature rose to 37.4° C. Express his temperature in Fahrenheit degrees.

_____

17.   Iron melts at 2,795° F. Gold melts at 1,063° C. Which of these metals has the higher melting point?

_____

By how many degrees?   _____

18.   Tom has to heat a liquid to 80° C for a chemistry experiment. His thermometer has a Fahrenheit scale only. To what Fahrenheit temperature reading should he heat the liquid?   _____

Kelvin Scale

Another temperature scale used with the metric system is called the Kelvin scale. It was named after Lord Kelvin, a great British physicist.

As shown below, the starting or zero point on the Kelvin scale is *absolute zero*. Absolute zero is the lowest theoretical temperature that a gas can reach.

Notice that the difference between the freezing and the boiling temperatures of water is 100 Celsius units and also 100 Kelvin units. The only difference between the two scales is that the Kelvin scale has a "head start" of 273.15 units.

You can change a Celsius reading (c) to a Kelvin reading (k) as follows.

$k = c + 273.15$

$72°C = \underline{\ ?\ }°K$
$k = 72 + 273.15$
$k = 345.15$
$72°C = 345.15°K$

You can change a Kelvin reading to a Celsius reading as follows.

$c = k - 273.15$

$250°K = \underline{\ ?\ }°C$
$c = 250 - 273.15$
$c = {}^-23.15$
$250°K = {}^-23.15°C$

By laying the edge of a ruler or paper perpendicular to the scales shown here, you can estimate equivalent temperatures on all three scales.

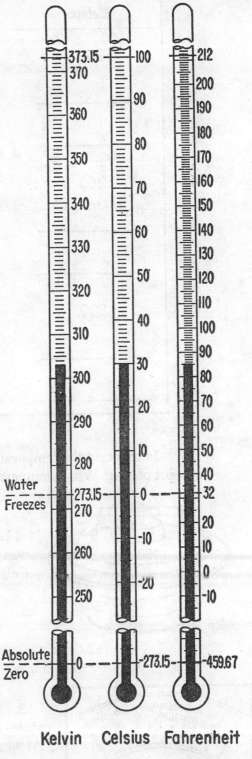

Kelvin   Celsius   Fahrenheit

Use the scales on page 103 to estimate the missing temperature readings.

	Kelvin	Celsius	Fahrenheit
1.	350	_____	_____
2.	_____	60	_____
3.	_____	_____	0
4.	260	_____	_____
5.	_____	⁻20	_____
6.	_____	_____	200
7.	300	_____	_____
8.	_____	90	_____
9.	_____	_____	50

10.   Jane found the temperature of some water to be 277°. The water was not boiling. Which temperature scale did she use? _____

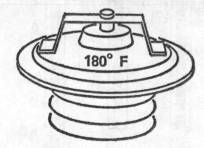

11.   George removed the old thermostat from his car. It was marked 180° F. New thermostats are marked in degrees Celsius. What marking should he look for on a new one?

_____

12.   In a physics experiment, the temperature of a gas was lowered to −413.5°. Which temperature scale was used? _____

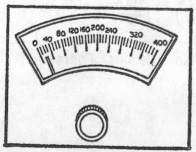

13.   The temperature control for a new oven used a Celsius scale. A cake recipe calls for an oven temperature of 350° F. At what Celsius reading should you set the control?

_____

Radian

In the standard system, angles are measured in degrees (a 360th part of the circumference of a circle). In the metric system, measuring angles in **radians** is preferred. Both scales are shown on the protractor below.

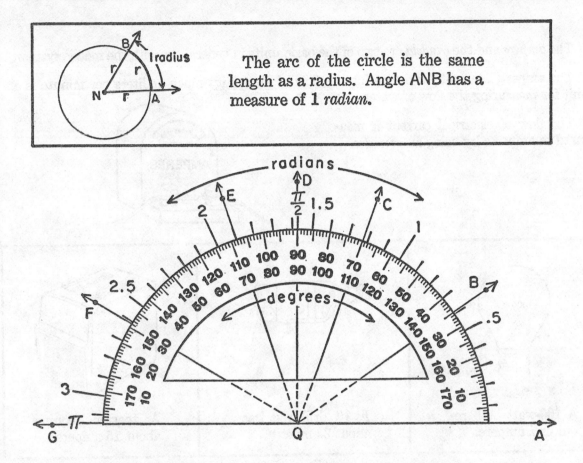

The arc of the circle is the same length as a radius. Angle ANB has a measure of 1 *radian*.

Use the scales shown with this protractor to find the approximate measures of each angle.

	angle	degrees	radians
1.	AQB	_____	_____
2.	AQC	_____	_____
3.	AQD	_____	_____
4.	AQE	_____	_____

Ampere, Candela, and Lumen

The *ampere* and the *candela* are two of the basic units in general use with the metric system.

The **ampere** is a unit for measuring the flow of electricity much as liters per minute is a unit for measuring the flow of water.

The flow of electrical current is measured in amperes by using an *ammeter*.

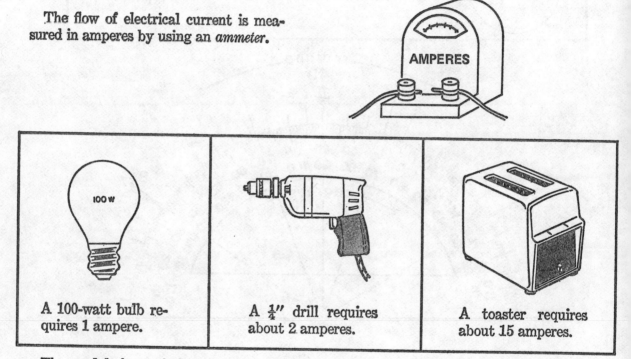

A 100-watt bulb requires 1 ampere.	A $\frac{1}{4}''$ drill requires about 2 amperes.	A toaster requires about 15 amperes.

The **candela** is a unit for measuring the luminous intensity (amount) of light produced by a light source.

The **lumen** is a unit for measuring the brightness of light when it reaches the surface of an object.

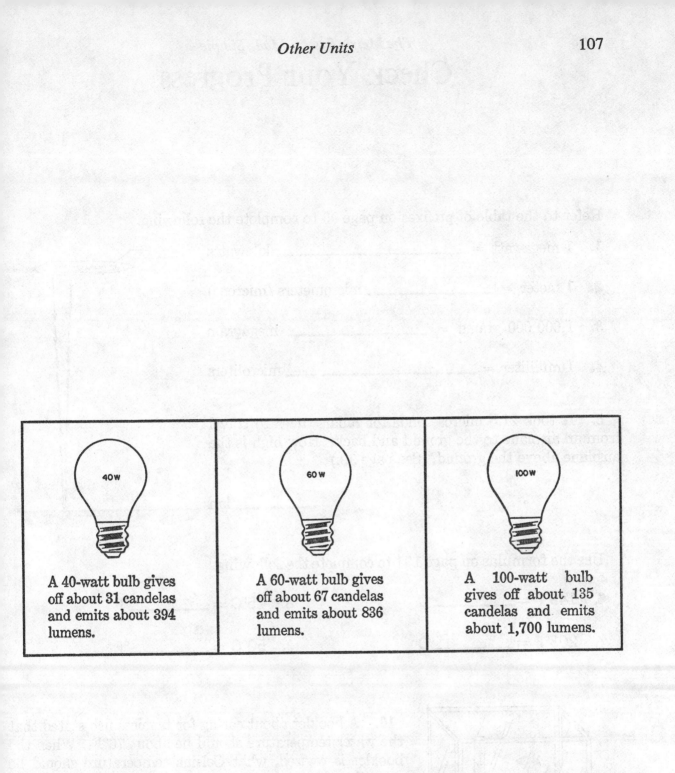

A 40-watt bulb gives off about 31 candelas and emits about 394 lumens.

A 60-watt bulb gives off about 67 candelas and emits about 836 lumens.

A 100-watt bulb gives off about 135 candelas and emits about 1,700 lumens.

Check Your Progress

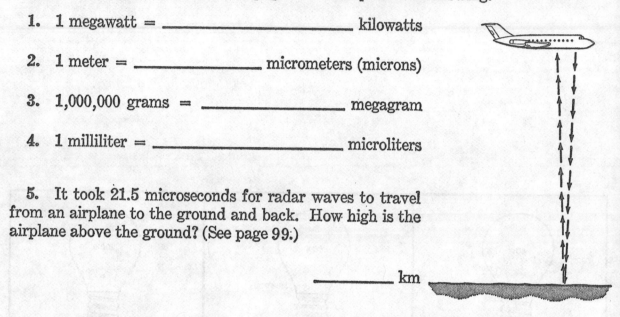

Refer to the table of prefixes on page 99 to complete the following.

1. 1 megawatt = _____ kilowatts

2. 1 meter = _____ micrometers (microns)

3. 1,000,000 grams = _____ megagram

4. 1 milliliter = _____ microliters

5. It took 21.5 microseconds for radar waves to travel from an airplane to the ground and back. How high is the airplane above the ground? (See page 99.)

_____ km

Use the formulas on page 101 to complete the following.

6. 149°F = _____ °C

7. 203°F = _____ °C

8. 15°C = _____ °F

9. 50°C = _____ °F

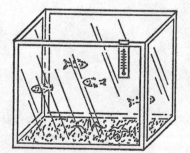

10. A booklet about caring for tropical fish stated that the water temperature should be about 76°F. When the booklet is revised, what Celsius temperature should be stated?

_____

Use the temperature scales on page 103 to estimate the missing temperature readings.

	Kelvin	Celsius	Fahrenheit
11.	_____	_____	175
12.	_____	5	_____
13.	295	_____	_____

Glossary

Ampere A unit for measuring the flow of electricity. Symbol: amp.

Area Amount of surface, measured in square units.

Are A metric surface measure, equal to 100 m². Symbol: a.

Atto- A prefix indicating one quintillionth of a given unit.

Barrel The amount contained in a barrel; especially the amount (as 31 gallons of fermented beverage or 42 gallons of petroleum) fixed for a certain commodity and used as a unit of measure for that particular commodity. Symbol: bbl.

Boardfoot A unit of quantity for lumber equal to the volume of a board $12 \times 12 \times 1$ inches. Symbol: fbm.

Bushel A unit of dry capacity equal to 4 pecks (2150.42 in³) or 35.238 liters

Candela A unit for measuring the luminous intensity (amount) of a light produced by a light source.

Capacity See Volume.

Celsius The name of the scale for temperature commonly used in conjunction with the metric system. Also known as the Centigrade scale. In the Celsius scale, water boils at 100° C and freezes at 0° C, as opposed to 212° F and 32° F, respectively, in the Fahrenheit scale. Symbol: ° C.

Centare A metric surface measure equal to 1 m². Symbol: ca.

Centi- A prefix indicating one hundredth of a given unit.

Centigram One hundredth of a gram. Symbol: cg.

Centiliter One hundredth of a liter. Symbol: cl.

Centimeter One hundredth of a meter. One centimeter equals .3937 inch. Symbol: cm.

Chain A unit of measure equal to 66 feet (20.1168 meters). Symbol: ch.

Cubic unit symbols Examples: mm³, cm³, m³, etc., used to denote volume.

Customary units Units of weights and measures currently in use in the United States, known also as English units. These include: inches, feet, yards, and miles for length; ounces, pounds, and tons for weight; pints, quarts, and gallons.

Deci- A prefix indicating one tenth of a given unit.

Decigram One tenth of a gram. Symbol: dg.

Deciliter One tenth of a liter. Roughly equal to .21 pint. Symbol: dl.

Decimeter Ten centimeters or one tenth of a meter. Symbol: dm.

Deka- A prefix indicating ten times a given unit.

Dekagram Ten grams. Symbol: dag.

Dekaliter Ten liters, roughly equivalent to 2.64 gallons. Symbol: dal.

Dekameter Ten meters. One dekameter roughly equals 10.91 yards. Symbol: dam.

Density The weight of any sample of a substance divided by the volume measure of that sample.

Dram A unit of avoirdupois weight equal to 27.343 grains or .0625 ounce (1.771 grams). Symbol: dr.

Fathom A unit of length equal to 6 feet (1.8288 meters) used for measuring the depth of water. Symbol: fath.

Femto- A prefix indicating one quadrillionth of a given unit.

Furlong A unit of distance equal to 220 yards (201.168 meters). No symbol.

Giga- A prefix indicating a billion times a given unit.

Gill A unit of liquid measure equal to .25 pint or 118.291 milliliters.

Grain A unit of weight equal to .002083 ounce (.0648 gram), originally based on the weight of a grain of wheat. Symbol: gr.

Gram A common metric unit of weight equal to one thousandth of a kilogram. Symbol: g.

Hectare The common unit of land measure in the metric system, equal to 100 ares or 10,000 square meters and equivalent to 2.471 acres. Symbol: ha.

Hecto- A prefix indicating one hundred times a given unit.

Hectogram One hundred grams. Symbol: hg.

Hectoliter One hundred liters. Symbol: hl.

Hectometer One hundred meters. Symbol: hm.

Hogshead A U.S. unit of capacity equal to 63 gallons (238.4809 liters). Symbol: hka.

Hundredweight A unit of weight (avoirdupois) commonly equivalent to 100 lbs. (45.359 kilograms) in the United States and 112 lbs (50.803 kilograms) in England. The former is known as the short hundredweight and the latter as the long hundredweight. Symbol: cwt.

Kelvin scale A temperature scale often used with the metric system and developed by the British physicist Lord Kelvin. The starting or zero point on the Kelvin scale is absolute zero ($-273.15°$ C, $-459.67°$ F) —the lowest theoretical temperature that a gas can reach. On this scale, water freezes at $273.15°$ K and boils at $373.15°$ K.

Kilo- A prefix indicating one thousand times a given unit.

Kilogram The standard unit of mass in the metric system. The kilogram is a cylinder of platinum-iridium alloy kept by the International Bureau of Weights and Measures near Paris. A duplicate kilogram is kept by the National Bureau of Standards in Washington and serves as the mass standard for the United States. One kilogram is approximately equal to 2.2 pounds. Symbol: kg.

Kiloliter One thousand liters. Symbol: kl.

Kilometer One thousand meters, equivalent to 3,280.8 feet or .621 mile. Symbol: km.

Link One of the standardized divisions of a surveyor's chain that is 7.92 inches (201.168 millimeters) long and serves as a measure of length. No symbol.

Liter The basic metric unit of liquid measure, equal to the volume of one kilogram of water at $4°$ C or one cubic decimeter. A liter is equivalent to 1.057 quarts. Symbol: l.

Lumen A unit for measuring the brightness of light when it reaches the surface of an object.

Mass The amount of material in an object, measured in kilograms (q.v.).

Mega- A prefix indicating one million times a given unit.

Meter The basic unit of length in the metric system. It is defined in terms of the wavelength of orange-red light emitted by a krypton-86 atom (1,650,763.73 such wavelengths to the meter). One meter equals 39.37 inches. Symbol: m.

Metric system A decimal system of weights and measures, adopted first in France and now in common use worldwide.

Metric ton One thousand kilograms, roughly equivalent to 2,200 pounds. Symbol: t.

Micron The millionth part of a meter. Symbol: μ.

Mile, International Nautical A unit of distance in sea and air navigation equal to 1.852 kilometers or 6,076.1033 feet.

Mill A unit of money (but not an actual coin) used primarily in accounting.

Milli- A prefix indicating one thousandth of a given unit.

Milligram One thousandth of a gram. Symbol: mg.

Milliliter One thousandth of a liter. Symbol: ml.

Millimeter One tenth of a centimeter or one thousandth of a meter. Symbol: mm.

Minim The smallest unit of liquid measure, the sixtieth part of a fluid dram, roughly equivalent to one drop.

Nano- A prefix indicating one billionth of a given unit.

Ounce, avoirdupois A unit of weight equal to 437.5 grains or .625 pound avoirdupois (28.349 grams). Symbol: oz. avdp.

Ounce, troy A unit of weight equal to 480 grains or .833 pound troy (31.103 grams). Symbol: oz. tr.

Peck A dry measure of 8 quarts or the fourth part of a bushel (8.89 liters).

Perimeter The measure of the distance around a figure.

Pico- A prefix indicating one trillionth of a given unit.

Pound, avoirdupois A unit of weight and mass equal to 7,000 grains (.453 kilogram) divided into 16 ounces, used for ordinary commercial purposes. Symbol: lb. avdp.

Pound, troy A unit of weight equal to 5,760 grains (.373 kilogram) divided into 12 ounces troy, used for gold, silver, and other precious metals. Symbol: lb. tr.

Radian An arc of a circle equal in length to the radius of that circle. An angle emanating from the center of a circle that subtends (cuts off) such an arc is said to measure one radian. Measuring angles in radians is preferred with the metric system.

Rod A unit of linear, 5.5 yards or 16.5 feet (5.0292 meters). A unit of surface measure 30.25 yd^2 (25.2901 m^2). No symbol.

Second The sixtieth part of a minute of a degree, often represented by the sign ″ as in 13 15′ 45″, read as 13 degrees, 15 minutes, 45 seconds.

Specific gravity The ratio of the density of a substance to the density of water at 4° C.

Square unit symbol Examples: mm^2, cm^2, m^2, etc.

Stere A cubic measure equivalent to 35.315 cubic feet or 1.3080 cubic yards (1.001 m^3). Used to measure cordwood. No symbol.

Tera- A prefix indicating a trillion times a given unit.

Ton, metric See Metric ton.

Volume The measure in cubic units of the amount of space inside any given container; also the measure of the amount such a container will hold. The latter is known as the *capacity* of the container and can be given in either units of liquid measure (see Liter, also Milliliter) or in cubic units.

Weight The force of the earth's pull on an object. Weight, in the Metric system, is commonly measured in grams.

Some Units and Their Symbols

UNIT	SYMBOL
ampere	amp
are	a
barrel	bbl
board foot	fbm
bushel	bu
candela	no symbol
carat	c
Celsius, degree	°C
centare	ca
centigram	cg
centiliter	cl
centimeter	cm
chain	ch
cubic centimeter	cm³
cubic decimeter	dm³
cubic dekameter	dam³
cubic hectometer	hm³
cubic kilometer	km³
cubic meter	m³
cubic millimeter	mm³
decigram	dg
deciliter	dl
decimeter	dm
dekagram	dag
dekaliter	dal
dekameter	dam
dram, avoirdupois	dr. avdp.
fathom	fath
furlong	no symbol
grain	no symbol
gram	g
hectare	ha
hectogram	hg
hectoliter	hl
hectometer	hm

hundredweight	cwt
International Nautical Mile	INM
Kelvin, degree	°K
kilogram	kg
kiloliter	kl
kilometer	km
link	no symbol
liter	l
lumen	no symbol
meter	m
microgram	μg
microliter	μl
micron	μ
milligram	mg
milliliter	ml
millimeter	mm
minim	no symbol
ounce	oz.
ounce, avoirdupois	oz. avdp.
ounce, troy	oz. tr.
peck	no symbol
pound	lb.
pound, avoirdupois	lb. avdp.
pound, troy	lb. tr.
radian	no symbol
rod	no symbol
second	s('')
square centimeter	cm²
square decimeter	dm²
square dekameter	dam²
square hectometer	hm²
square kilometer	km²
square meter	m²
square millimeter	mm²
stere	no symbol
ton, metric	t

Units of Measurement—Conversion Factors*

Units of Length

To Convert from Centimeters

To	Multiply by
Inches _ _ _ _ _ _ _ _ _ _	0.393 700 8
Feet _ _ _ _ _ _ _ _ _ _ _	0.032 808 40
Yards _ _ _ _ _ _ _ _ _ _ _	0.010 936 13
Meters _ _ _ _ _ _ _ _ _ _ _	**0.01**

To Convert from Meters

To	Multiply by
Inches _ _ _ _ _ _ _ _ _ _	39.370 08
Feet _ _ _ _ _ _ _ _ _ _ _	3.280 840
Yards _ _ _ _ _ _ _ _ _ _ _	1.093 613
Miles _ _ _ _ _ _ _ _ _ _ _	0.000 621 37
Millimeters _ _ _ _ _ _ _ _	**1,000**
Centimeters _ _ _ _ _ _ _ _	**100**
Kilometers _ _ _ _ _ _ _ _	**0.001**

To Convert from Inches

To	Multiply by
Feet _ _ _ _ _ _ _ _ _ _ _	0.083 333 33
Yards _ _ _ _ _ _ _ _ _ _	0.027 777 78
Centimeters _ _ _ _ _ _ _ _	**2.54**
Meters _ _ _ _ _ _ _ _ _ _	**0.025 4**

To Convert from Feet

To	Multiply by
Inches _ _ _ _ _ _ _ _ _ _	**12**
Yards _ _ _ _ _ _ _ _ _ _	0.333 333 3
Miles _ _ _ _ _ _ _ _ _ _	0.000 189 39
Centimeters _ _ _ _ _ _ _ _	**30.48**
Meters _ _ _ _ _ _ _ _ _ _	**0.304 8**
Kilometers _ _ _ _ _ _ _ _	**0.000 304 8**

To Convert from Yards

To	Multiply by
Inches _ _ _ _ _ _ _ _ _ _	**36**
Feet _ _ _ _ _ _ _ _ _ _ _	**3**
Miles _ _ _ _ _ _ _ _ _ _	0.000 568 18
Centimeters _ _ _ _ _ _ _ _	**91.44**
Meters _ _ _ _ _ _ _ _ _ _	**0.914 4**

To Convert from Miles

To	Multiply by
Inches _ _ _ _ _ _ _ _ _ _	**63,360**
Feet _ _ _ _ _ _ _ _ _ _ _	**5,280**
Yards _ _ _ _ _ _ _ _ _ _	**1,760**
Centimeters _ _ _ _ _ _ _ _	**160,934.4**
Meters _ _ _ _ _ _ _ _ _ _	**1,609.344**
Kilometers _ _ _ _ _ _ _ _	**1.609 344**

* All boldface figures are exact; the others generally are given to seven significant figures.

In using conversion factors, it is possible to perform division as well as the multiplication process shown here. Division may be particularly advantageous where more than the significant figures published here are required. Division may be performed in lieu of multiplication by using the reciprocal of any indicated multiplier as divisor. For example, to convert from centimeters to inches by division, refer to the table headed "To Convert from *Inches*" and use the factor listed at "centimeters" (*2.54*) as divisor.

Units of Mass

To Convert from Grams

To	Multiply by
Grains	15.432 36
Avoirdupois drams	0.564 383 4
Avoirdupois ounces	0.035 273 96
Troy ounces	0.032 150 75
Troy pounds	0.002 679 23
Avoirdupois pounds	0.002 204 62
Milligrams	1,000
Kilograms	0.001

To Convert from Kilograms

To	Multiply by
Grains	15,432.36
Avoirdupois drams	564.383 4
Avoirdupois ounces	35.273 96
Troy ounces	32.150 75
Troy pounds	2.679 229
Avoirdupois pounds	2.204 623
Grams	1,000
Short hundredweights	0.022 046 23
Short tons	0.001 102 31
Long tons	0.000 984 2
Metric tons	0.001

To Convert from Metric Tons

To	Multiply by
Avoirdupois pounds	2,204.623
Short hundredweights	22.046 23
Short tons	1.102 311 3
Long tons	0.984 206 5
Kilograms	1,000

To Convert from Grains

To	Multiply by
Avoirdupois drams	0.036 571 43
Avoirdupois ounces	0.002 285 71
Troy ounces	0.002 083 33
Troy pounds	0.000 173 61
Avoirdupois pounds	0.000 142 86
Milligrams	64.798 91
Grams	0.064 798 91
Kilograms	0.000 064 798 91

To Convert from Avoirdupois Pounds

To	Multiply by
Grains	7,000
Avoirdupois drams	256
Avoirdupois ounces	16
Troy ounces	14.583 33
Troy pounds	1.215 278
Grams	453.592 37
Kilograms	0.453 592 37
Short hundredweights	0.01
Short tons	0.000 5
Long tons	0.000 446 428 6
Metric tons	0.000 453 592 37

To Convert from Avoirdupois Ounces

To	Multiply by
Grains	437.5
Avoirdupois drams	16
Troy ounces	0.911 458 3
Troy pounds	0.075 954 86
Avoirdupois pounds	0.062 5
Grams	28.349 523 125
Kilograms	0.028 349 523 125

To Convert from Short Hundredweights

To	Multiply by
Avoirdupois pounds	100
Short tons	0.05
Long tons	0.044 642 86
Kilograms	45.359 237
Metric tons	0.045 359 237

To Convert from Short Tons

To	Multiply by
Avoirdupois pounds	2,000
Short hundredweights	20
Long tons	0.892 857 1
Kilograms	907.184 74
Metric tons	0.907 184 74

To Convert from
Troy Ounces

To	Multiply by
Grains _ _ _ _ _ _ _ _ _ _ _	480
Avoirdupois drams _ _ _ _ _ _	17.554 29
Avoirdupois ounces _ _ _ _ _ _	1.097 143
Troy pounds _ _ _ _ _ _ _ _	0.083 333 3
Avoirdupois pounds _ _ _ _ _ _	0.068 571 43
Grams _ _ _ _ _ _ _ _ _ _	31.103 476 8

To Convert from
Troy Pounds

To	Multiply by
Grains _ _ _ _ _ _ _ _ _ _	5,760
Avoirdupois drams _ _ _ _ _	210.651 4
Avoirdupois ounces _ _ _ _ _	13.165 71
Troy ounces _ _ _ _ _ _ _	12
Avoirdupois pounds _ _ _ _ _	0.822 857 1
Grams _ _ _ _ _ _ _ _ _ _ _	373.241 721 6

To Convert from
Long Tons

To	Multiply by
Avoirdupois ounces _ _ _	35,840
Avoirdupois pounds _ _ _	2,240
Short hundredweights _ _	22.4
Short tons _ _ _ _ _ _ _	1.12
Kilograms _ _ _ _ _ _ _	1,016.046 908 8
Metric tons _ _ _ _ _ _	1.016 046 908 8

Units of Capacity, or Volume, Liquid Measure

To Convert from Milliliters

To	Multiply by
Minims	16.230 73
Liquid ounces	0.033 814 02
Gills	0.008 453 5
Liquid pints	0.002 113 4
Liquid quarts	0.001 056 7
Gallons	0.000 264 17
Cubic inches	0.061 023 74
Liters	0.001

To Convert from Gills

To	Multiply by
Minims	1,920
Liquid ounces	4
Liquid pints	0.25
Liquid quarts	0.125
Gallons	0.031 25
Cubic inches	7.218 75
Cubic feet	0.004 177 517
Milliliters	118.294 118 25
Liters	0.118 294 118 25

To Convert from Cubic Meters

To	Multiply by
Gallons	264.172 05
Cubic inches	61,023.74
Cubic feet	35.314 67
Liters	1,000
Cubic yards	1.307 950 6

To Convert from Liquid Ounces

To	Multiply by
Minims	480
Gills	0.25
Liquid pints	0.062 5
Liquid quarts	0.031 25
Gallons	0.007 812 5
Cubic inches	1.804 687 5
Cubic feet	0.001 044 38
Milliliters	29.573 53
Liters	0.029 573 53

To Convert from Liters

To	Multiply by
Liquid ounces	33.814 02
Gills	8.453 506
Liquid pints	2.113 376
Liquid quarts	1.056 688
Gallons	0.264 172 05
Cubic inches	61.023 74
Cubic feet	0.035 314 67
Milliliters	1,000
Cubic meters	0.001
Cubic yards	0.001 307 95

To Convert from Cubic Inches

To	Multiply by
Minims	265.974 0
Liquid ounces	0.554 112 6
Gills	0.138 528 1
Liquid pints	0.034 632 03
Liquids quarts	0.017 316 02
Gallons	0.004 329 0
Cubic feet	0.000 578 7
Milliliters	16.387 064
Liters	0.016 387 064
Cubic meters	0.000 016 387 064
Cubic yards	0.000 021 43

To Convert from Minims

To	Multiply by
Liquid ounces	0.002 083 33
Gills	0.000 520 83
Milliliters	0.061 611 52
Cubic inches	0.003 759 77

To Convert from Liquid Pints

To	Multiply by
Minims	7,680
Liquid ounces	16
Gills	4
Liquid quarts	0.5
Gallons	0.125
Cubic inches	28.875
Cubic feet	0.016 710 07
Milliliters	473.176 473
Liters	0.473 176 473

To Convert from Liquid Quarts

To	Multiply by
Minims	15,360
Liquid ounces	32
Gills	8
Liquid pints	2
Gallons	0.25
Cubic inches	57.75
Cubic feet	0.033 420 14
Milliliters	946.352 946
Liters	0.946 352 946

To Convert from Cubic Feet

To	Multiply by
Liquid ounces	957.506 5
Gills	239.376 6
Liquid pints	59.844 16
Liquid quarts	29.922 08
Gallons	7.480 519
Cubic inches	1,728
Liters	28.316 846 592
Cubic meters	0.028 316 846 592
Cubic yards	0.037 037 04

To Convert from Gallons

To	Multiply by
Minims	61,440
Liquid ounces	128
Gills	32
Liquid pints	8
Liquid quarts	4
Cubic inches	231
Cubic feet	0.133 680 6
Milliliters	3,785.411 784
Liters	3.785 411 784
Cubic meters	0.003 785 411 784
Cubic yards	0.004 951 13

To Convert from Cubic Yards

To	Multiply by
Gallons	201.974 0
Cubic inches	46,656
Cubic feet	27
Liters	764.554 857 984
Cubic meters	0.764 554 857 984

Units of Capacity, or Volume, Dry Measure

To Convert from
Liters

To	Multiply by
Dry pints	1.816 166
Dry quarts	0.908 082 98
Pecks	0.113 510 4
Bushels	0.028 377 59
Dekaliters	0.1

To Convert from
Dry Quarts

To	Multiply by
Dry pints	2
Pecks	0.125
Bushels	0.031 25
Cubic inches	67.200 625
Cubic feet	0.038 889 25
Liters	1.101 221
Dekaliters	0.110 122 1

To Convert from
Dekaliters

To	Multiply by
Dry pints	18.161 66
Dry quarts	9.080 829 8
Pecks	1.135 104
Bushels	0.283 775 9
Cubic inches	610.237 4
Cubic feet	0.353 146 7
Liters	10

To Convert from
Pecks

To	Multiply by
Dry pints	16
Dry quarts	8
Bushels	0.25
Cubic inches	537.605
Cubic feet	0.311 114
Liters	8.809 767 5
Dekaliters	0.880 976 75
Cubic meters	0.008 809 77
Cubic yards	0.011 522 74

To Convert from
Cubic Meters

To	Multiply by
Pecks	113.510 4
Bushels	28.377 59

To Convert from
Dry Pints

To	Multiply by
Dry quarts	0.5
Pecks	0.062 5
Bushels	0.015 625
Cubic inches	33.600 312 5
Cubic feet	0.019 444 63
Liters	0.550 610 47
Dekaliters	0.055 061 05

To Convert from
Bushels

To	Multiply by
Dry pints	64
Dry quarts	32
Pecks	4
Cubic inches	2,150.42
Cubic feet	1.244 456
Liters	35.239 07
Dekaliters	3.523 907
Cubic meters	0.035 239 07
Cubic yards	0.046 090 96

To Convert from Cubic Inches	
To	Multiply by
Dry pints _ _ _ _ _ _ _ _ _ _	0.029 761 6
Dry quarts _ _ _ _ _ _ _ _ _	0.014 880 8
Pecks _ _ _ _ _ _ _ _ _ _ _	0.001 860 10
Bushels _ _ _ _ _ _ _ _ _ _	0.000 465 025

To Convert from Cubic Yards	
To	Multiply by
Pecks _ _ _ _ _ _ _ _ _ _ _	86.784 91
Bushels _ _ _ _ _ _ _ _ _ _	21.696 227

To Convert from Cubic Feet	
To	Multiply by
Dry pints _ _ _ _ _ _ _ _ _	51.428 09
Dry quarts _ _ _ _ _ _ _ _	25.714 05
Pecks _ _ _ _ _ _ _ _ _ _	3.214 256
Bushels _ _ _ _ _ _ _ _ _	0.803 563 95

Units of Area

To Convert from Square Centimeters

To	Multiply by
Square inches	0.155 000 3
Square feet	0.001 076 39
Square yards	0.000 119 599
Square meters	0.000 1

To Convert from Square Feet

To	Multiply by
Square inches	144
Square yards	0.111 111 1
Acres	0.000 022 957
Square centimeters	929.030 4
Square meters	0.092 903 04

To Convert from Square Meters

To	Multiply by
Square inches	1,550.003
Square feet	10.763 91
Square yards	1.195 990
Acres	0.000 247 105
Square centimeters	10,000
Hectares	0.000 1

To Convert from Square Yards

To	Multiply by
Square inches	1,296
Square feet	9
Acres	0.000 206 611 6
Square miles	0.000 000 322 830 6
Square centimeters	8,361.273 6
Square meters	0.836 127 36
Hectares	0.000 083 612 736

To Convert from Hectares

To	Multiply by
Square feet	107,639.1
Square yards	11,959.90
Acres	2.471 054
Square miles	0.003 861 02
Square meters	10,000

To Convert from Acres

To	Multiply by
Square feet	43,560
Square yards	4,840
Square miles	0.001 562 5
Square meters	4,046.856 422 4
Hectares	0.404 685 642 24

To Convert from Square Inches

To	Multiply by
Square feet	0.006 944 44
Square yards	0.000 771 605
Square centimeters	6.451 6
Square meters	0.000 645 16

To Convert from Square Miles

To	Multiply by
Square feet	27,878,400
Square yards	3,097,600
Acres	640
Square meters	2,589,988.110 336
Hectares	258.998 811 033 6

Special Tables

Equivalents of Decimal and Binary Fractions of an Inch in Millimeters

From 1/64 Inch to 1 Inch

½'s	¼'s	8ths	16ths	32ds	64ths	Milli-meters	Decimals of an inch	Inch	½'s	¼'s	8ths	16ths	32ds	64ths	Milli-meters	Decimals of an inch
					1	= 0.397	0.015625							33	= 13.097	0.515625
				1	2	= .794	.03125						17	34	= 13.494	.53125
					3	= 1.191	.046875							35	= 13.891	.546875
			1	2	4	= 1.588	.0625					9	18	36	= 14.288	.5625
					5	= 1.984	.078125							37	= 14.684	.578125
				3	6	= 2.381	.09375						19	38	= 15.081	.59375
					7	= 2.778	.109375							39	= 15.478	.609375
		1	2	4	8	= 3.175	.1250				5	10	20	40	= 15.875	.625
					9	= 3.572	.140625							41	= 16.272	.640625
				5	10	= 3.969	.15625						21	42	= 16.669	.65625
					11	= 4.366	.171875							43	= 17.066	.671875
			3	6	12	= 4.762	.1875					11	22	44	= 17.462	.6875
					13	= 5.159	.203125							45	= 17.859	.703125
				7	14	= 5.556	.21875						23	46	= 18.256	.71875
					15	= 5.953	.234375							47	= 18.653	.734375
	1	2	4	8	16	= 6.350	.2500			3	6	12	24	48	= 19.050	.75
					17	= 6.747	.265625							49	= 19.447	.765625
				9	18	= 7.144	.28125						25	50	= 19.844	.78125
					19	= 7.541	.296875							51	= 20.241	.796875
			5	10	20	= 7.938	.3125					13	26	52	= 20.638	.8125
					21	= 8.334	.328125							53	= 21.034	.828125
				11	22	= 8.731	.34375						27	54	= 21.431	.84375
					23	= 9.128	.359375							55	= 21.828	.859375
		3	6	12	24	= 9.525	.3750				7	14	28	56	= 22.225	.875
					25	= 9.922	.390625							57	= 22.622	.890625
				13	26	= 10.319	.40625						29	58	= 23.019	.90625
					27	= 10.716	.421875							59	= 23.416	.921875
			7	14	28	= 11.112	.4375					15	30	60	= 23.812	.9375
					29	= 11.509	.453125							61	= 24.209	.953125
				15	30	= 11.906	.46875						31	62	= 24.606	.96875
					31	= 12.303	.484375							63	= 25.003	.984375
1	2	4	8	16	32	= 12.700	.5	1	2	4	8	16	32	64	= 25.400	1.000

International Nautical Miles and Kilometers

Basic relation: International Nautical Mile = 1.852 kilometers

Int. Nautical Miles	Kilometers	Int. Nautical Miles	Kilometers	Int. Nautical Miles	Kilometers
0		40	74.080	80	148.160
1	1.852	1	75.932	1	150.012
2	3.704	2	77.784	2	151.864
3	5.556	3	79.636	3	153.716
4	7.408	4	81.488	4	155.568
5	9.260	5	83.340	5	157.420
6	11.112	6	85.192	6	159.272
7	12.964	7	87.044	7	161.124
8	14.816	8	88.896	8	162.976
9	16.668	9	90.748	9	164.828
10	18.520	50	92.600	90	166.680
1	20.372	1	94.452	1	168.532
2	22.224	2	96.304	2	170.384
3	24.076	3	98.156	3	172.236
4	25.928	4	100.008	4	174.088
5	27.780	5	101.860	5	175.940
6	29.632	6	103.712	6	177.792
7	31.484	7	105.564	7	179.644
8	33.336	8	107.416	8	181.496
9	35.188	9	109.268	9	183.348
20	37.040	60	111.120	100	185.200
1	38.892	1	112.972		
2	40.744	2	114.824		
3	42.596	3	116.676		
4	44.448	4	118.528		
5	46.300	5	120.380		
6	48.152	6	122.232		
7	50.004	7	124.084		
8	51.856	8	125.936		
9	53.708	9	127.788		
30	55.560	70	129.640		
1	57.412	1	131.492		
2	59.264	2	133.344		
3	61.116	3	135.196		
4	62.968	4	137.048		
5	64.820	5	138.900		
6	66.672	6	140.752		
7	68.524	7	142.604		
8	70.376	8	144.456		
9	72.228	9	146.308		

Kilometers	Int. Nautical Miles	Kilometers	Int. Nautical Miles	Kilometers	Int. Nautical Miles
0		40	21.5983	80	43.1965
1	0.5400	1	22.1382	1	43.7365
2	1.0799	2	22.6782	2	44.2765
3	1.6199	3	23.2181	3	44.8164
4	2.1598	4	23.7581	4	45.3564
5	2.6998	5	24.2981	5	45.8963
6	3.2397	6	24.8380	6	46.4363
7	3.7797	7	25.3780	7	46.9762
8	4.3197	8	25.9179	8	47.5162
9	4.8596	9	26.4579	9	48.0562
10	5.3996	50	26.9978	90	48.5961
1	5.9395	1	27.5378	1	49.1361
2	6.4795	2	28.0778	2	49.6760
3	7.0194	3	28.6177	3	50.2160
4	7.5594	4	29.1577	4	50.7559
5	8.0994	5	29.6976	5	51.2959
6	8.6393	6	30.2376	6	51.8359
7	9.1793	7	30.7775	7	52.3758
8	9.7192	8	31.3175	8	52.9158
9	10.2592	9	31.8575	9	53.4557
				100	53.9957
20	10.7991	60	32.3974		
1	11.3391	1	32.9374		
2	11.8790	2	33.4773		
3	12.4190	3	34.0173		
4	12.9590	4	34.5572		
5	13.4989	5	35.0972		
6	14.0389	6	35.6371		
7	14.5788	7	36.1771		
8	15.1188	8	36.7171		
9	15.6587	9	37.2570		
30	16.1987	70	37.7970		
1	16.7387	1	38.3369		
2	17.2786	2	38.8769		
3	17.8186	3	39.4168		
4	18.3585	4	39.9568		
5	18.8985	5	40.4968		
6	19.4384	6	41.0367		
7	19.9784	7	41.5767		
8	20.5184	8	42.1166		
9	21.0583	9	42.6566		

Selected Tables of
Equivalents

Length

Inches to Millimeters
1 in = 25.4 mm

IN	MM
.50	12.700
1.00	25.400
1.50	38.100
2.00	50.800
2.50	63.500
3.00	76.200
3.50	88.900
4.00	101.600
4.50	114.300
5.00	127.000
5.50	139.700
6.00	152.400
6.50	165.100
7.00	177.800
7.50	190.500
8.00	203.200
8.50	215.900
9.00	228.600
9.50	241.300
10.00	254.000
10.50	266.700
11.00	279.400
11.50	292.100
12.00	304.800

Millimeters to Inches
1 mm = .0393700787 in

MM	IN
.50	.01969
1.00	.03937
1.50	.05906
2.00	.07874
2.50	.09843
3.00	.11811
3.50	.13780
4.00	.15748
4.50	.17717
5.00	.19685
5.50	.21654
6.00	.23622
6.50	.25591
7.00	.27559
7.50	.29528
8.00	.31496
8.50	.33465
9.00	.35433
9.50	.37402
10.00	.39370
20.00	.78740
30.00	1.18110
40.00	1.57480
50.00	1.96850

Feet to Meters
1 ft = .3048 m

FT	M
1	.3048
2	.6096
3	.9144
4	1.2192
5	1.5240
6	1.8288
7	2.1336
8	2.4384
9	2.7432
10	3.0480
11	3.3528
12	3.6576
13	3.9624
14	4.2672
15	4.5720
20	6.0960
30	9.1440
40	12.1920
50	15.2400
60	18.2880
70	21.3360
80	24.3840
90	27.4320
100	30.4800

Meters to Feet
1 m = 3.280839895 ft

M	FT
1	3.2808
2	6.5617
3	9.8425
4	13.1234
5	16.4042
6	19.6850
7	22.9659
8	26.2467
9	29.5276
10	32.8084
11	36.0892
12	39.3701
13	42.6509
14	45.9318
15	49.2126
20	65.6168
30	98.4252
40	131.2336
50	164.0420
60	196.8504
70	229.6588
80	262.4672
90	295.2756
100	328.0840

Yards to Meters
1 yd = .9144 m

YD	M
1	.9144
2	1.8288
3	2.7432
4	3.6575
5	4.5720
6	5.4864
7	6.4008
8	7.3152
9	8.2296
10	9.1440
11	10.0584
12	10.9728
13	11.8872
14	12.8016
15	13.7160
20	18.2880
30	27.4320
40	36.5760
50	45.7200
60	54.8640
70	64.0080
80	73.1520
90	82.2960
100	91.4400

Meters to Yards
1 m = 1.093613298 yds

M	YD
1	1.0936
2	2.1872
3	3.2808
4	4.3754
5	5.4681
6	6.5617
7	7.6553
8	8.7489
9	9.8425
10	10.9361
11	12.0297
12	13.1234
13	14.2170
14	15.3106
15	16.4042
20	21.8723
30	32.8084
40	43.7445
50	54.6807
60	65.6168
70	76.5529
80	87.4891
90	98.4252
100	109.3613

Miles to Kilometers
1 mi = 1.609344 km

MI	KM
1	1.6903
2	3.2187
3	4.8280
4	6.4374
5	8.0467
6	9.6561
7	11.2654
8	12.8748
9	14.4841
10	16.0934
11	17.7028
12	19.3121
13	20.9215
14	22.5308
15	24.1402
20	32.1869
30	48.2803
40	64.3738
50	80.4672
60	96.5606
70	112.6541
80	128.7475
90	144.8410
100	160.9344

Kilometers to Miles
1 km = .621371192 mi

KM	MI
1	.62137
2	1.24274
3	1.86411
4	2.48548
5	3.10686
6	3.72823
7	4.34960
8	4.97097
9	5.59234
10	6.21371
11	6.83508
12	7.45645
13	8.07783
14	8.69920
15	9.32057
20	12.42742
30	18.64114
40	24.85485
50	31.06856
60	37.28227
70	43.49598
80	49.70970
90	55.92341
100	62.13712

Mass

Grains to Milligrams

1 grain = 64.79891 mg

GRAINS	MG
1	64.80
2	129.60
3	194.40
4	259.20
5	323.99
6	388.79
7	453.59
8	518.39
9	583.19
10	647.99
11	712.79
12	777.59
13	842.39
14	907.18
15	971.98
20	1295.98
30	1943.97
40	2591.96
50	3239.95
60	3887.93
70	4535.92
80	5183.91
90	5831.90
100	6479.89

Milligrams to Grains

1 mg = .01543235835 grain

MG	GRAINS
1	.015432
2	.030865
3	.046297
4	.061729
5	.077162
6	.092594
7	.108027
8	.123459
9	.138891
10	.154324
11	.169756
12	.185188
13	.200621
14	.216053
15	.231485
20	.308647
30	.46297
40	.617294
50	.771618
60	.925941
70	1.080265
80	1.234589
90	1.388912
100	1.543236

Avoirdupois Ounces to Grams

1 oz. avdp. = 28.349523125 g

OZ. AVDP.	G
1	28.350
2	56.699
3	85.049
4	113.398
5	141.748
6	170.097
7	198.447
8	226.796
9	255.146
10	283.495
11	311.845
12	340.194
13	368.544
14	396.893
15	425.243
16	453.592
20	566.990
30	850.486
40	1133.981
50	1417.476
60	1700.971
70	1984.467
80	2267.962
90	2551.457
100	2834.952

Grams to Avoirdupois Ounces

1 g = .035273962 oz. avdp.

G	OZ. AVDP.
1	.035274
2	.070548
3	.105822
4	.141096
5	.176370
6	.211644
7	.246918
8	.282192
9	.317466
10	.352740
11	.388014
12	.423288
13	.458562
14	.493835
15	.529109
20	.705479
30	1.058219
40	1.410958
50	1.763698
60	2.116438
70	2.469177
80	2.821917
90	3.174657
100	3.527396

Avoirdupois Pounds to Kilograms

1 lb. avdp. = .45359237 kg

Kilograms to Avoirdupois Pounds

1 kg = 2.204622622 lb. avdp.

LB. AVDP.	KG	KG	LB. AVDP.
1	.45359	1	2.2046
2	.90718	2	4.4092
3	1.36078	3	6.6139
4	1.81537	4	8.8185
5	2.26796	5	11.0231
6	2.72155	6	13.2277
7	3.17515	7	15.4324
8	3.62874	8	17.6370
9	4.80233	9	19.8416
10	4.53592	10	22.0462
11	4.98952	11	24.2508
12	5.44311	12	26.4555
13	5.89670	13	28.6601
14	6.35029	14	30.8647
15	6.80389	15	33.0693
20	9.07185	20	44.0925
30	13.60777	30	66.1387
40	18.14369	40	88.1849
50	22.67962	50	110.2311
60	27.21554	60	132.2774
70	31.75147	70	154.3236
80	36.28739	80	176.3698
90	40.82331	90	198.4160
100	45.35924	100	220.4623
110	49.89516	110	242.5085
120	54.43108	120	264.5547
130	58.96701	130	286.6009
140	63.50293	140	308.6472
150	68.03886	150	330.6934
160	72.57478	160	352.7396
170	77.11070	170	374.7858
180	81.64663	180	396.8321
190	86.18255	190	418.8783
200	90.71847	200	440.9245
210	95.25440	210	462.9707
220	99.79032	220	485.0170
230	104.32625	230	507.0632
240	108.86217	240	529.1094
250	113.39809	250	551.1557
260	117.93402	260	573.2019
270	122.46994	270	595.2481
280	127.00586	280	617.2943
290	131.54179	290	639.3406
300	136.07771	300	661.3868

Capacity, or Volume

Cubic Inches to Milliliters
$1\ in^3 = 16.387064\ ml$

IN³	ML
1	16.387
2	32.774
3	49.161
4	65.548
5	81.935
6	98.322
7	114.709
8	131.097
9	147.484
10	163.871
11	180.258
12	196.645
13	213.032
14	229.419
15	245.806
20	327.741
30	491.612
40	655.483
50	819.353
60	983.224
70	1147.094
80	1310.965
90	1474.836
100	1638.706

Milliliters to Cubic Inches
$1\ ml = .061023744\ in^3$

ML	IN³
1	.061024
2	.122047
3	.13071
4	.244095
5	.305119
6	.366142
7	.427166
8	.488190
9	.549214
10	.610237
11	.671261
12	.732285
13	.793309
14	.854332
15	.915356
20	1.220475
30	1.830712
40	2.440950
50	3.051187
60	3.661425
70	4.271662
80	4.881899
90	5.492137
100	6.102374

Cubic Feet to Cubic Meters
$1\ ft^3 = .028316846592\ m^3$

FT³	M³
1	.028317
2	.056634
3	.084951
4	.113267
5	.141584
6	.169901
7	.198218
8	.226535
9	.254852
10	.283168
11	.311485
12	.339802
13	.368119
14	.396436
15	.424753
20	.566337
30	.849505
40	1.132674
50	1.415842
60	1.699011
70	1.982179
80	2.265348
90	2.548516
100	2.831685

Cubic Meters to Cubic Feet
$1\ m^3 = 35.3146667\ ft^3$

M³	FT³
1	35.315
2	70.629
3	105.944
4	141.259
5	176.573
6	211.888
7	247.203
8	282.517
9	317.832
10	353.147
11	388.461
12	423.776
13	459.091
14	494.405
15	529.720
20	706.293
30	1059.440
40	1412.587
50	1765.733
60	2118.880
70	2472.027
80	2825.173
90	3178.320
100	3531.467

Cubic Yards to Cubic Meters

1 yd^3 = .764554857984 m^3

YD^3	M^3
1	.76455
2	1.52911
3	2.29366
4	3.05822
5	3.82277
6	4.58733
7	5.35188
8	6.11644
9	6.88099
10	7.64555
11	8.41010
12	9.17466
13	9.93921
14	10.70377
15	11.46832
20	15.29110
30	22.93665
40	30.58219
50	38.22774
60	45.87329
70	53.51884
80	61.16439
90	68.80994
100	76.45549

Cubic Meters to Cubic Yards

1 m^3 = 1.307950619 yd^3

M^3	YD^3
1	1.3080
2	2.6159
3	3.9239
4	5.2318
5	6.5398
6	7.8477
7	9.1557
8	10.4636
9	11.77.6
10	13.0795
11	14.3875
12	15.6954
13	17.0034
14	18.3113
15	19.6193
20	26.1590
30	39.2385
40	52.3180
50	65.3975
60	78.4770
70	91.5565
80	104.6360
90	117.7156
100	130.7951

Liquid Ounces to Milliliters

1 oz. liq. = 29.5735296 ml

OZ. LIQ.	ML
1	29.574
2	59.147
3	88.721
4	118.294
5	147.868
6	177.441
7	207.015
8	236.588
9	266.162
10	295.735
11	325.309
12	354.882
13	384.456
14	414.020
15	443.603
20	591.471
30	887.206
40	1182.941
50	1478.676
60	1774.412
70	2070.147
80	2365.882
90	2661.618
100	2957.353

Milliliters to Liquid Ounces

1 ml = .03381402 oz. liq.

ML	OZ. LIQ.
1	.03381
2	.06763
3	.10144
4	.13526
5	.16907
6	.20288
7	.23670
8	.27051
9	.30433
10	.33814
11	.37195
12	.40577
13	.43958
14	.47340
15	.50721
20	.67628
30	1.01442
40	1.53256
50	1.69070
60	2.02884
70	2.36698
80	2.70512
90	3.04326
100	3.38140

Liquid Quarts to Liters

1 qt. liq. = .946352949 l

QT. LIQ.	L
1	.94635
2	1.89271
3	2.83906
4	3.78541
5	4.73176
6	5.67812
7	6.62447
8	7.57082
9	8.51718
10	9.46353
11	10.40988
12	11.35624
13	12.30259
14	13.24894
15	14.19529
20	18.92706
30	28.39059
40	37.85412
50	47.31765
60	56.78118
70	66.24471
80	75.70823
90	85.17176
100	94.63529

Liters to Liquid Quarts

1 l = 1.0566882 qt. liq.

L	QT. LIQ.
1	1.0567
2	2.1134
3	3.1701
4	4.2268
5	5.2834
6	6.3401
7	7.3968
8	8.4535
9	9.5102
10	10.5669
11	11.6236
12	12.6803
13	13.7369
14	14.7936
15	15.8503
20	21.1338
30	31.7006
40	42.2675
50	52.8344
60	63.4013
70	73.9682
80	84.5351
90	95.1019
100	105.6688

Gallons to Liters

1 gal = 3.785411784 l

GAL	L
1	3.7854
2	7.5708
3	11.3562
4	15.1416
5	18.9271
6	22.7125
7	26.4979
8	30.2833
9	34.0687
10	37.8541
11	41.6395
12	45.4249
13	49.2104
14	52.9958
15	56.7812
20	75.7082
30	113.5624
40	151.4165
50	189.2706
60	227.1247
70	264.9788
80	302.8329
90	340.6871
100	378.5412

Liters to Gallons

1 l = .264172052 gal

L	GAL
1	.26417
2	.52834
3	.79252
4	1.05669
5	1.32086
6	1.58503
7	1.84920
8	2.11338
9	2.37755
10	2.64172
11	2.90589
12	3.17006
13	3.43424
14	3.69841
15	3.96258
20	5.28344
30	7.92516
40	10.56688
50	13.20860
60	15.85032
70	18.49204
80	21.13376
90	23.77548
100	26.41721

Area

Square Inches to Square Centimeters
1 in² = 6.4516 cm²

IN²	CM²
1	6.4516
2	12.9032
3	19.3548
4	25.8064
5	32.2580
6	38.7096
7	45.1612
8	51.6128
9	58.0644
10	64.5160
11	70.9676
12	77.4192
13	83.8708
14	90.3224
15	96.7740
20	129.0320
30	193.5480
40	258.0640
50	322.5800
60	387.0960
70	451.6120
80	516.1280
90	580.6440
100	645.1600

Square Centimeters to Square Inches
1 cm² = .15500031 in²

CM²	IN²
1	.15500
2	.31000
3	.46500
4	.62000
5	.77500
6	.93000
7	1.08500
8	1.24000
9	1.39500
10	1.55000
11	1.70500
12	1.86000
13	2.01500
14	2.17000
15	2.32500
20	3.10001
30	4.65001
40	6.20001
50	7.75002
60	9.30002
70	10.85002
80	12.40002
90	13.95003
100	15.50003

Square Feet to Square Meters
1 ft² = .09290304 m²

FT²	M²
1	.09290
2	.18581
3	.27871
4	.37161
5	.46452
6	.55742
7	.65032
8	.74322
9	.83613
10	.92903
11	1.02193
12	1.11484
13	1.20774
14	1.30064
15	1.39355
20	1.85806
30	2.78709
40	3.71612
50	4.64515
60	5.57418
70	6.50321
80	7.43224
90	8.36127
100	9.29030

Square Meters to Square Feet
1 m² = 10.7639104 ft²

M²	FT²
1	10.764
2	21.528
3	32.292
4	43.056
5	53.820
6	64.583
7	75.347
8	86.111
9	96.875
10	107.639
11	118.403
12	129.167
13	139.931
14	150.695
15	161.459
20	215.278
30	322.917
40	430.556
50	538.196
60	645.835
70	753.474
80	861.113
90	968.752
100	1076.391

Square Yards to Square Meters
1 yd² = .83612736 m²

YD²	M²
1	.83613
2	1.67225
3	2.50838
4	3.34451
5	4.18064
6	5.01676
7	5.85289
8	6.68902
9	7.52515
10	8.36127
11	9.19740
12	10.03353
13	10.86966
14	11.70578
15	12.54191
20	16.72255
30	25.08382
40	33.44509
50	41.80637
60	50.16764
70	58.52892
80	66.89019
90	75.25146
100	83.61274

Square Meters to Square Yards
1 m² = 1.195990046 yd²

M²	YD²
1	1.1960
2	2.3920
3	3.5880
4	4.7840
5	5.9800
6	7.1759
7	8.3719
8	9.5679
9	10.7639
10	11.9599
11	13.1559
12	14.3519
13	15.5479
14	16.7439
15	17.9399
20	23.9198
30	35.8797
40	47.8396
50	59.7995
60	71.7594
70	83.7193
80	95.6792
90	107.6391
100	119.5990

Acres to Hectares
1 ac = .40468564224 ha

AC	HA
1	.40469
2	.80937
3	1.21406
4	1.61874
5	2.02343
6	2.42811
7	2.83280
8	3.23749
9	3.64217
10	4.04686
11	4.45154
12	4.85623
13	5.26019
14	5.66560
15	6.07028
20	8.09371
30	12.14057
40	16.18743
50	20.23428
60	24.28114
70	28.32799
80	32.37485
90	36.42171
100	40.46856

Hectares to Acres
1 ha = 2.4710538 ac

HA	AC
1	2.4711
2	4.9421
3	7.4132
4	9.8842
5	12.3553
6	14.8263
7	17.2974
8	19.7684
9	22.2395
10	24.7105
11	27.1816
12	29.6526
13	32.1237
14	34.5948
15	37.0658
20	49.4211
30	74.1316
40	98.8422
50	123.5527
60	148.2632
70	172.9738
80	197.6843
90	222.3948
100	247.1054

Answers

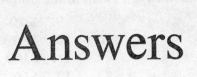

Pages 1–2
1. 6
2. 2
3. 2
4. meter
5. 100
6. 91–92
7. 8–9
8–11. Answers will vary.

Pages 3–4
1. 10
2. 4
3. 9
4. 10
5. 2.5
6. 1.9
7. 6
8. 2.4
9. .1
10. 1.5; 15
11. 10; 2.63; 26.3
12. 15; 1.5
13. .1; 26.3; 2.63
14. 60
15. 4.25
16. 12
17. 8
18. .62
19. .05
20. 41
21. 2.9
22. 500
23. 6
24. 528

Pages 5–6
1. 2.5
2. 25–26

3. 10
4. 10
5. 6
6. 62
7. 9
8. 89
9. 5
10. 52
11. 91–92
12. 914–15
13. 25.5
14. 254–55
15. 19.5
16. 193
17. 3.25; 32.5
18. 10; 512
19. .1; 6; .6
20. .1; 1.25; .125
21. 170
22. .24
23. 4.5
24. 5
25. 6.2
26. 420
27. .013
28. .75
29. 12
30. $.001
31. 1 cent = 10 mills and 1 cm = 10 mm

Pages 7–8
1. 10
2. 100
3. 1
4. 10; 4.2; 42
5. .1; 50; 5
6. 10
7. 1

8. 1; 109.4
9. 2; 218.8
10. 4; 437.6
11. 8; 875.2
12. 15; 1,641.0
13. 20
14. 40
15. 80
16. 150

Pages 9–10
1. 3,168
2. 1,000
3. mile
4. 6
5. meter
6. kilometer
7. 5,000
8. 250
9. 6,400
10. 3
11. .412
12. .1
13. Mary; 128 km
14. San Francisco through Houston to New York; 4,904 km
15. 296 km
16. 74 km/hour
17. 260 km
18. 4 hr
19. 320 km
20. 484 km
21. 3.5 hr

Pages 11–13
1. Latin
2. Greek
3. Greek
4. Latin
5. Greek
6. Latin
7. $1,000
8. $.001

9. $.10
10. $.01
11. $100
12. $10
13. 4th = 40 h = 400 t = 4,000 ones = 40,000 ts
14. 4 km = 40 hm = 400 dam = 4,000 m = 40,000 dm
15. 400 t = 4,000 ones = 40,000ts = 400,000 hs = 4,000,000 ths
16. 400 dam = 4,000 m = 40,000 dm = 400,000 cm = 4,000,000 mm
17. .02 h = .2 t = 2 ones = 20 ts = 200 hs
18. .02 hm = .2 dam = 2 m = 20 dm = 200 cm
19. 10
20. 10
21. 10
22. .1
23. .1
24. .1
25. 1.3
26. .002
27. .04
28. 100
29. 100
30. 100
31. .01
32. .01
33. .01
34. .002
35. 14
36. 35
37. 1,000
38. .001
39. 10,000
40. 2,000,000
41. .0002
42. .002
43. .003
44. 3,000
45. .00003

Pages 14–16

1. Estimates may vary.
2. 6
3. Estimates may vary; 8
4. Estimates may vary; 2
5. Estimates may vary; 6
6. Estimates may vary; 48
7. Estimates may vary; 35
8. Estimates may vary; 70
9–12. Student drawings.

(Estimates will vary.)

13. 7
14. 5
15. 3
16. 8
17. 6
18. 11
19. 3
20. 9
21. 36; 3, 6
22. 45; 4, 5
23. 42; 4, 2
24. 65; 6, 5
25. 18; 1, 8
26. 113; 11, 3
27. 93; 9, 3
28. 28; 2, 8
29–31. Answers will vary.

Pages 17–19

1. 43
2. 2, 4
3. 1, 9
4. 81
5. 54
6. 80
7. 6 cm 1 mm
8. 9 cm 2 mm
9. 7 cm 7 mm
10. 63; 6, 3
11. 53; 5, 3
12. 93; 9, 3
13. 5 m 9 dm 8 cm
14. 7 m 6 dm 1 cm
15. 8 m 1 dm 1 cm
16. 8 dm 9 cm 1 mm
17. 7 dm 1 cm 3 mm
18. 7 dm 3 cm 3 mm
19. 6 dam 9 m 9 dm
20. 7 dam 9 m 1 dm
21. 9 dam 2 m 3 dm
22. 5 hm 3 dam
23. 8 hm 3 dam 7 m
24. 7 km 2 hm 5 dam

Pages 20–21

1. 7, 8
2. No; no
3. 12
4. 15 cm
5. 19 mm
6. 19 m
7. 5 cm 1 mm
8. 2 cm 7 mm
9. 8 mm
10. 3 dm 7 cm
11. 1 m 9 dm
12. 7 m 8 dm
13. 4 dm 2 cm 3 mm
14. 4 dm 2 cm 5 mm
15. 2 dm 7 cm 2 mm
16. 4 dm 5 cm 8 mm
17. 4 m 6 dm 7 cm
18. 3 m 0 dm 7 cm
19. 3 dam 8 m 2 dm
20. 1 dam 5 m 8 dm
21. 3 hm 1 dam 8 m
22. 2 hm 5 dam 8 m
23. 2 km 9 hm 2 dam
24. 2 km 5 hm 7 dam

Pages 22–23

1. 3, 5
2. 5, 7

3. 9, 2
4. 4, 6
5. 2, 8
6. 7, 4
7. ant; 1, 8
8. 1, 3
9. 1, 6
10. 3
11. 14, 9
12. 7
13. 6
14. 12
15. 38
16. 76
17. 3, 0, 9
18. 3, 7
19. 9, 8
20. 2, 8, 8

Pages 24–27

1. 7, 3
2. 6, 6
3. first; 7
4. library; 3
5. 738
6. 82
7. 542
8. 7, 5
9. 1, 9
10. 4,690
11. 30
12. 790
13. 11,180
14. 6,980
15. 8
16. 16
17. 35
18. 30
19. 90
20. 90
21. 5,400
22. 60; 180; 180; 10,800
23. 152,456; 456; 27,360

24. 19.5 mm
25. 17.5 mm
26. 1.40 mm; 1.10 mm
27. 1.65 mm; 1.45 mm

Pages 28–31

1. 3
2. 48
3. 31
4. 5
5. 31
6. 8
7. 2
8. 8
9. 3½
10. 18
11. 28
12. 89
13. 48
14. 4.8
15. 50
16. 500
17. 24
18. 240
19. 350
20. 120
21. 3,200
22. 70
23. 20
24. 1,000
25. 17
26. 250
27. 320
28. 47
29. 33
30. 33
31. 36
32. 36
33. .33
34. 36
35. 78–79; .78–.79
36. 90; 35.5
37. 1,560; 15.6

38. 850; 335
39. 405; 4.05
40. 14; 46
41. 125; 12,500
42. 15; 1,500
43. 16; 52.5
44. 29; 2,900

Pages 32–34

1. 11
2. 8
3. 5
4. 71
5. 94
6. 115
7. 6; 63
8. 5; 47
9. 10; 96
10. .01; hundredths
11. deci; tenths
12. deka; 10
13. 100; hundreds
14. kilo; thousands
15. 10
16. 100
17. 1,000
18. 10
19. 10
20. 10
21. 100
22. 1,000
23. 10
24. 10
25. 40
26. 300
27. 500
28. .014
29. .0235
30. 1.42
31. 5 m 8 dm 1 cm
32. 7 dm 3 cm 4 mm
33. 4 m 6 dm 3 cm
34. 1 dm 8 cm 4 mm

35. 8, 2
36. 5, 6
37. 88 km/hr
38. 770 km
39. bus; 122 km
40. 153 km
41. Yes
42. 28 mm

Pages 35–38

1. Estimates may vary; 12
2. Estimates may vary; 14
3. Estimates may vary; 12
4. Estimates may vary; 80
5. Estimates may vary; 130
6. Estimates may vary; 140
7. Estimates may vary; 92
8. Estimates may vary; 133
9. 212
10. 15, 6
11. 73, 2
12. 21, 8
13. 16, 7
14. 5, 1
15. 16
16. 3, 5
17. 6, 7
18. **24 m**
19. 1 m; 2.2 m

Pages 39–42

1. mm^2
2. dm^2
3. 10
4. 100
5. 10
6. 100
7. sq mm
8. sq dm
9. 100
10. 100
11. greater

12. less
13. 6 or 7
14. 6
15. 15
16. 8
17. 4
18. 3
19. 12
20. 5
21. 3
22. 15
23. 2.5
24. 4.5
25. 11.25
26. 20
27. 20
28. 400
29. 15
30. 30
31. 450
32. 10
33. 25
34. 250
35. 120.35
36. 5.04
37. 1,849
38. 7.65 cm²

Pages 43–44

1. 4
2. 6
3. 24
4. 4
5. 6
6. 24
7. 24
8. 588
9. 550
10. 54
11. 391
12. 5,888
13. 252,000
14. 319

15. 19
16. 15

Pages 45–47

1. $A = s^2$
2. $A = b \times h$
3. $A = \frac{1}{2} h (a + b)$
4. $A = \frac{1}{2} b \times h$
5. $A = \pi \times r^2$
6. $A = 1 \times w$
7. 81
8. 1,008
9. 12.56
10. 3.04
11. 65.28
12. 250
13. .0192
14. 6
15. 28.26
16. 1.5; 1; 1.5
17. 1; 3.14
18. 1; 3; 1; 2
19. 1.5; 2.25
20. 2; 1; 1
21. .5; .785
22. 3; 1; 3
23. 4; 1.5; 6
24. 3
25. 12
26. 4
27. 1
28. 4
29. 16
30. The area is 4 times greater.

Pages 48–50

1. 1
2. 100
3. 10,000
4. 100
5. 100
6. 10,000
7. Larger

8. 4,900
9. 98
10. Larger
11. 2,400 ca; 24 a
12. 300 ca; 3 a
13. 48 ca; .48 a
14. 1,525
15. 1,830
16. 2,440
17. 40
18. B & C
19. 25
20. 50
21. 25
22. 12.5

Page 51
1. No
2. The square

Page 52
1. 12
2. 1.2
3. 35
4. 350
5. 4.4
6. 24
7. 64
8. 220
9. 120
10. 36
11. 40
12. 50
13. 750
14. 1,200
15. 160
16. 220

Pages 53–54
1. 23
2. 32
3. 11, 8
4. 26

5. 840
6. 729
7. 391
8. 696
9. .0196
10. 50.24
11. 112
12. 7
13. 5
14. 40
15. 750
16. 200
17. 2

Pages 55–56
1. mm³
2. dm³
3. cc
4. 10
5. 100
6. 1,000
7. 10
8. 100
9. 1,000
10. Larger
11. Smaller

Pages 57–59
1. 4
2. 3
3. 12
4. 12
5. 2
6. 24
7. 24
8. 24
9. 24
10. 120
11. 54
12. 11,250
13. 10.648
14. 44,640

15. 13.125
16. 400
17. 105
18. 20
19. 9,600
20. 70
21. 18,400

Pages 60–61
1. 1
2. 10
3. 1,000
4. 1,000,000
5. 12,600
6. 240
7. 270
8. 1,392
9. 24; 24,000
10. 189
11. 9
12. 750

Pages 62–63
1. 27
2. 54
3. 8
4. 12
5. 6
6. 1
7. No
8. 2 by 3 by 4
9. 1 by 1 by 24

Pages 64–67
1. Same amount
2. 1,000
3. 1
4. Slightly more
5. 1,000
6. About 946
7. Yes
8. No
9. No, 1 liter is more than 1 quart.
10. 1

11. 7,540
12. 456
13. 65,000
14. 500
15. 3.45
16. .0785
17. .005
18. .25
19. 1
20. 1
21. 5
22. 5
23. 48
24. 240
25. 240
26. 180
27. 1,000
28. 10
29. 10,000
30. 10,000
31. 2.64
32. About 3,788
33. 1,000
34. 100
35. 10
36. 1/10
37. 1/100
38. 1/1000
39. C
40. 300

Page 68
1. 1,000
2. 1,000
3. 1
4. 3,000
5. 3,000
6. 3
7. 12
8. 12

Pages 69–72
1. 270 m³
2. $324

3. 15,726.41
4. 15.72641
5. 1,450 ml; 1.45 liters
6. $.59
7. 400
8. 750 ml; 250 ml
9. 520
10. 1; 36
11. 31
12. 720
13. 400
14. 15,014.4; 15.0144
15. 466
16. 10
17. 463
18. 6.5
19. 5.8135
20. No
21. 30
22. 25
23. 15.8
24. 26.25

Pages 73–74
1. 65
2. 6.5
3. 35
4. 3.5
5. 11
6. 150
7. 182
8. 123
9. 270
10. 38.5
11. 53
12. 950
13. .19
14. 8.9
15. 180
16. 19

Pages 75–76
1. 120
2. 600

3. 540
4. 64
5. 1
6. 1,000
7. 1,000
8. 1
9. 10
10. 50
11. 9
12. 12
13. 4,000
14. 5
15. 4,800
16. .0765
17. 2.5
18. 7
19. 3
20. 30,000
21. 4,000
22. 4,000
23. 4
24. 20
25. 5

Pages 78–81
1. More
2. 2.2
3. 2.2
4. .0022
5. 3,000
6. 50
7. 2.5
8. .0355
9. 100
10. 5
11. 5,000
12. 100,000
13. 90,000
14. 500
15. .725
16. .02
17. 200
18. No
19. 1,000

20. 1
21. 13
22. 22
23. 55
24. 9
25. 36
26. 16
27. Tom; 1 kilogram is more than 1 pound
28. 25; 25

Pages 82–86
1. 454
2. 454; 28.375
3. 907
4. Less
5. 1,000,000
6. .815
7. .185
8. 1,000
9. 185
10. 185; .185
11. 100 kg (.1 t)
12. 1
13. 1
14. 1
15. 6,500
16. 6,500,000
17. 750
18. 3,000
19. 2
20. 5,000
21. 2.5
22. .0035
23. .0145
24. 1
25. 1,000
26. 1,000
27. 1
28. 1,000
29. 15
30. 280
31. 4,500

32. 4.5
33. 72.5
34. .5
35. 750
36. .75
37. 3,375
38. About 1.95
39. 121.68
40. 54,872
41. 54,872
42. 54.872
43. Metric units; the computation is much simpler.
44. 22
45. 16
46. Less, 97
47. 45
48. Less

Pages 87–92
1. 1 g
2. 1 mg
3. 1 kg
4. 10 g
5. 2 kg
6. 10 kg
7. 1 kg
8. 80 kg
9. 60; 60; 60
10. 87
11. 1 kg is more than 2 pounds
12. 161
13. Janice
14. 45
15. 340.2; 326.9
16. 13.3 kg
17. C; 3.7 kg
18. E; 1.4 kg
19. 85.003 g or 85,003 mg
20. 2,785.003 g or 2,785,003 mg
21. 70.785003 kg or 70,785.003 g
22. Banana; 1.8 g
23. 4

24. 1⅓
25. 22.7 g (.05 lb); 68.1 g (.15 lb)
26. 4.99; 4,990
27. $.0025
28. $.002
29. 20,000
30. 1,500 g; 1.5 kg
31. 3,400.5 g; 3.4005 kg
32. 2,500.5 g (2.5005 kg)
33. 1,400; 1.4; 1,400; 1.4
34. 66.5
35. 4.95
36. 555.45
37. 93.55 kg
38. 931 kg
39. About 71 kg or 102 liters
40. 26,275; 26.275
41. 63,000; 63
42. 450
43. 350; .35; 350; .35
44. 46.8
45. 546.8 g
46. 475

Pages 93–95
1. 1
2. 1
3. 1
4. 3.5
5. .7
6. .7
7. 1 cc water; .3 g
8. Water; water is heavier than gasoline
9. 2.7
10. 19.3
11. .92
12. 7.9
13. 3.2
14. Gasoline, ice, water, aluminum, limestone, iron, gold
15. 1,000; 1,000
16. 700; 700
18. 3.5; 3,500 g

19. 1.6; 1,600 g
20. 1.7; 1,700 g
21. 8.9; 8,900 g
22. 110
23. 20
24. 5.5

Page 96
1. 35
2. 350
3. 32
4. 3.2
5. 7
6. 90
7. 395
8. .21
9. 11.4
10. 180
11. 365
12. .41
13. 4,650
14. 570
15. 1,980
16. 154

Pages 97–98
1. 4.5
2. 50
3. 1,000
4. 4,000
5. 1.5
6. 1,000
7. 7,000
8. 75,000
9. .5
10. 3,000
11. 650
12. 1,000
13. 200
14. 200
15. 200
16. 5
17. 1,000; 1

18. 80; 140
19. 600
20. 50½
21. 24.59055; 24,590.55
22. $.33
23. 296; 1.89; 1.36; 226.7; 113.3; .946; 473; 3.78; 453

Pages 99–100

1. 1,000,000
2. .000001
3. 1,000
4. 1,000,000
5. .186
6. About .3
7. About 13
8. About 280
9. 37.02
10. $60,000
11. 40,000; 40
12. 10

Page 102

1. 392
2. 59
3. 239
4. 113
5. 60
6. 35
7. 40
8. 55
9. 95
10. 45
11. 248
12. 25
13. 100° C
14. 2° C
15. 120° F
16. About 99.3° F
17. Iron; 472° C (849° F)
18. 176° F

Page 104

	KELVIN	CELSIUS	FAHRENHEIT
1.	350	77	170
2.	333	60	140
3.	255	—18	0
4.	260	—13	8
5.	253	—20	—4
6.	366	93	200
7.	300	27	80
8.	363	90	194
9.	283	10	50

10. Kelvin
11. About 82° C
12. Fahrenheit
13. About 177–80°

Page 105

1. 35; .61
2. 70; 1.22
3. 90; 1.57
4. 108; 1.88

Page 108

1. 1,000
2. 1,000,000
3. 1
4. 1,000
5. About 3.2
6. 65
7. 95
8. 59
9. 122
10. About 24° C

	KELVIN	CELSIUS	FAHRENHEIT
11.	353	79	175
12.	278	5	41
13.	295	22	71